D0436885

Six Sigma for Marketing Processes

Six Sigma for Marketing Processes

An Overview for Marketing Executives, Leaders, and Managers

Clyde M. Creveling
Lynne Hambleton
Burke McCarthy

Prentice Hall

Upper Saddle River, NJ • Boston • Indianapolis • San Francisco

New York • Toronto • Montreal • London • Munich • Paris • Madrid

Capetown • Sydney • Tokyo • Singapore • Mexico City

The publisher offers excellent discounts on this book when ordered in quantity for bulk purchases or special sales, which may include electronic versions and/or custom covers and content particular to your business, training goals, marketing focus, and branding interests. For more information, please contact

U.S. Corporate and Government Sales
800-382-3419
corpsales@pearsontechgroup.com

For sales outside the U.S., please contact

International Sales
international@pearsoned.com

This Book Is Safari Enabled

The Safari® Enabled icon on the cover of your favorite technology book means the book is available through Safari Bookshelf. When you buy this book, you get free access to the online edition for 45 days. Safari Bookshelf is an electronic reference library that lets you easily search thousands of technical books, find code samples, download chapters, and access technical information whenever and wherever you need it.

To gain 45-day Safari Enabled access to this book:

- Go to http://www.prenhallprofessional.com/safarienabled
- Complete the brief registration form
- Enter the coupon code 8PI5-AAEK-7J63-M9KT-P2L7

If you have difficulty registering on Safari Bookshelf or accessing the online edition, please e-mail customer-service@safaribooksonline.com

Visit us on the web: www.prenhallprofessional.com

ISBN 0-13-199008-X

Text printed in the United States on recycled paper at R. R. Donnelley & Sons in Crawfordsville, IN.

First printing February 2006

Library of Congress Cataloging-in-Publication Data

Creveling, Clyde M., 1956-

Six sigma for marketing processes : an overview for marketing executives, leaders, and managers / Clyde M. Creveling, Lynne Hambleton, Burke McCarthy.

p. cm.

ISBN 0-13-199008-X

1. Marketing—Management. 2. Marketing—Quality control. 3. Six sigma (Quality control standard) I. Hambleton, Lynne. II. McCarthy, Burke. III. Title.

HF5415.13.C74 2006

658.8'02—dc22

2005030766

PRENTICE HALL SIX SIGMA FOR INNOVATION AND GROWTH SERIES

Clyde (Skip) M. Creveling, Editorial Advisor,
Product Development Systems & Solutions Inc.

MARKETING PROCESSES

Six Sigma for Marketing Processes
Creveling, Hambleton, and McCarthy

TECHNICAL PROCESSES

Design for Six Sigma in Technology and Product Development
Creveling, Slutsky, and Antis

Tolerance Design
Creveling

PRENTICE HALL SIX SIGMA FOR INNOVATION AND GROWTH SERIES

What Six Sigma has been and is becoming has stimulated an exciting, new body of knowledge. The old form of Six Sigma is all about finding and fixing problems using the ubiquitous DMAIC process. Cost savings and defect reduction are its goals. Financial returns from DMAIC projects occur at the bottom line of the financial ledger. Many would agree that the DMADV process was the next logical step in the evolution of Six Sigma methodologies when we need to design a new business process. The five step models have served us well, but it is time to look into the future.

The new form of Six Sigma uses tools, methods and best practices which introduces an approach to efficiently produce growth results within your company's existing business processes. It is not focused on reactive problem-solving, but rather on prevention of problems during the work you do to innovate, refresh and invigorate your business. Its financial return is at the top-line of the financial ledger. Its goal—innovation leading to growth.

Many are asking what is the future of Six Sigma? Has it run its course? Our experience tells us that Six Sigma is alive, evolving and expanding to meet new market demands. This is why we are introducing a new series of books that seeks to communicate a newly emerging branch of Six Sigma that focuses on creativity and new business growth. The **Prentice Hall Six Sigma for Innovation and Growth Series** contains books in two general process arenas: Marketing processes and Technical processes. The books in this series will span strategic, tactical, and operational process arenas to transcend the ongoing activities within a business. Product Portfolio Renewal and R&D are strategic processes; Product and Service Commercialization are tactical processes and Post-Launch Product Line Management, Sales, Customer Service and Support as well as Technical Production, Technical Service, and Support Engineering are operational processes. These processes would benefit from the rigor and discipline that Six Sigma-enhanced work produces.

Expect great things from this new series of books if you are looking for ideas on how to improve innovation and growth on a sustainable basis. They will take you to the next level of learning and doing through Six Sigma enablement within your organization. Classic Six Sigma is serving us well on the cost-side and we see only good news on the horizon through the evolution of Six Sigma for Innovation and Growth!

Clyde (Skip) M. Creveling, Editorial Advisor,
Product Development Systems & Solutions Inc.

Skip Creveling: *I would like to dedicate this text to my lovely wife, Kathy, and my son, Neil. I would also like to dedicate it to my good friend and marketing colleague, Scott McGregor. Without his influence, insight, and globetrotting adventures in developing Six Sigma concepts in the world of marketing, this book simply would not exist.*

Lynne Hambleton: *I would like to dedicate my work in this book to Bill Magee and Lois Markt for their gentle, steadfast confidence and love, and to Shirley Edwards for her generous and sage guidance, all of which were crucial gifts that sparked energy and conviction. I also would like to thank coauthors Skip and Burke for their open and honest partnership and budding friendship.*

Burke McCarthy: *I dedicate my work in this book to Janet Crawford and Thomas and Megan Maeve.*

Contents

PREFACE

What Is In This Book?

This is not a book about marketing theory or basic marketing principles—we assume you are a marketing professional and know a good bit about marketing science fundamentals. This book is all about Six Sigma for marketing professionals. The kind of Six Sigma we explore is relatively new. It is the form of Six Sigma that focuses on

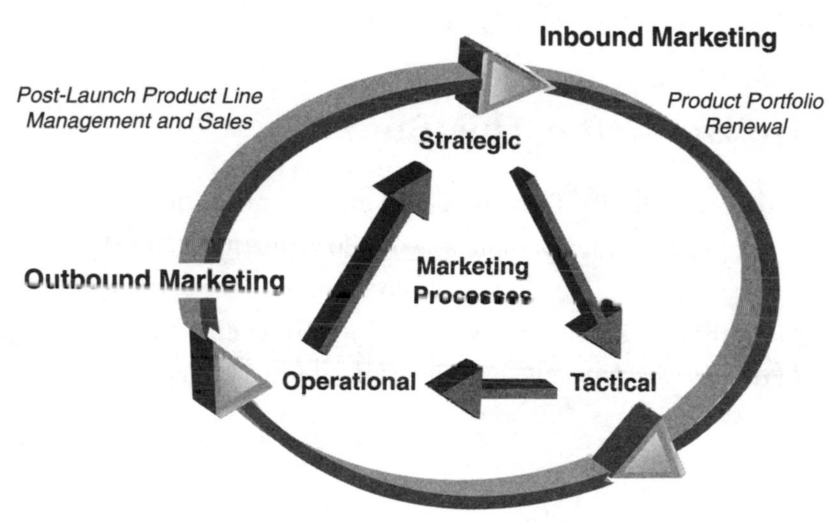

growth—that prevents problems by designing and structuring Six Sigma Sigma-enhanced work within marketing processes. Its boundaries encompass marketing's three process arenas for enabling a business to attain a state of sustainable growth.

1. **Strategic Marketing Process**—product or service portfolio renewal
2. **Tactical Marketing Process**—product or service commercialization
3. **Operational Marketing Process**—post-launch product or service line management

This book is reasonably short and is primarily intended as an overview for marketing executives, leaders, and managers. Anyone interested in the way Six Sigma tools, methods, and best practices enhance and enable these three marketing processes can benefit from this book. This book guides the reader in structuring a lean work flow for completing the right marketing tasks using the right tools, methods, and best practices—at the right time within the aforementioned processes. Yes, this book is all about Lean Six Sigma-enabled marketing.

Why We Wrote This Book

Why did we write it? To help take marketing professionals into the same kind of Six Sigma paradigm, work flow, measurement rigor, and lean process discipline that exists in the world of Design for Six Sigma (DFSS). Our first book, *Design for Six Sigma in Technology and Product Development* (Prentice Hall, 2003), has become a strong standard for research and development (R&D), product commercialization, and manufacturing support engineers. It is all about what to do and when to do it in the phases of technology development and product commercialization for engineering teams and their leaders. Every time we teach and mentor engineering teams on DFSS, they ask, "Where are the marketing people? Shouldn't they be here working with us as a team as we develop this new product?" The answer of course is always yes. So, a strong, new trend is occurring all over the world. It is a new form of collaborative innovation between those who practice DFSS and those who are beginning to practice Six Sigma for Marketing (SSFM). Two very harmonious bodies of Six Sigma knowledge are aligning and integrating into what we call Six Sigma for Innovation and Growth. In fact, this book is part of an exciting new series from Prentice Hall called the *Six Sigma for Innovation and Growth Series: Marketing Processes and Technical Processes*.

DFSS and SSFM are integrating to form a unified approach for those who are commercializing products together. This book, in part, is "DFSS for Marketing Professionals." We go far beyond simply talking about product commercialization in this book. We set the stage for a comprehensive Six Sigma-enabled work flow for marketing professionals. That work flow crosses the three process arenas we mentioned earlier—portfolio renewal (strategic in-bound marketing), commercialization (tactical in-bound marketing), and product or service line management (operational out-bound marketing). That is why the logo for this book looks the way it does. Take a moment to reflect on that image and you will see our view of the way marketing work flow is structured in the text.

About the Chapters

The book is laid out in eight chapters. Chapter 1, "Introduction to Six Sigma for Marketing Processes," presents the whole integrated story of Six Sigma in Marketing Processes. It covers the big picture of the way all three marketing process arenas work in harmony. One without the others is insufficient for actively sustaining growth in a business.

Chapters 2, "Measuring Marketing Performance and Risk Accrual Using Scorecards," and 3, "Six Sigma-Enabled Project Management in Marketing Processes," work closely together. Chapter 2 is about a system of integrated marketing scorecards that measure risk accrual from tool use to task completion to gate deliverables for any of the three marketing processes. Chapter 3 is a great way to get a project management view of how marketing teams can design and manage their work with a little help from some very useful Six Sigma tools (Monte Carlo Simulations and Project Failure Modes & Effects Analysis [FMEA]). Chapter 3 can help you lean out your marketing tasks and assess them for cycle-time risk.

Chapters 4," Six Sigma in the Strategic Marketing Process," 5," Six Sigma in the Tactical Marketing Process," and 6," Six Sigma in the Operational Marketing Process," contain more detailed views within each marketing process. The chapters lay out the gate requirements and gate deliverables within phase tasks and the enabling tools, methods, and best practices that help marketing teams complete their critical tasks. They offer a standard work set (a lean term) that can be designed into your marketing processes where you live on a daily basis. These chapters help you design your marketing work so you have efficient work flow and low variability in your summary results. This helps prevent problems and ultimately sustain growth. This is so because what you do adds value and helps assure your business cases reach their full entitlement. When business cases deliver what they promise—you will grow.

Chapter 7, "Quick Review of Traditional DMAIC," provides a brief overview of the important classic Six Sigma problem-solving approach known as Define-Measure-Analyze-Improve-Control (DMAIC).

Chapter 8, "Future Trends in Six Sigma and Marketing Processes," wraps everything up quickly and succinctly. We know marketing professionals are very busy folks, so we try to get the right information to you in a few short chapters so you can help lead your teams to new performance levels as you seek to sustain growth in your business.

A Word About Six Sigma Tools, Methods, and Best Practices

Six Sigma tools, methods, and best practices are in order at the outset of this book. When we discuss the various flows of marketing tasks, we find many opportunities to add value to them with well-known, time-tested combinations from Six Sigma (DMAIC, as well as DFSS). The following list helps set the stage for aligning marketing work with the numerous value-adding tools, methods, and best practices from Six Sigma. Once again, the difference this book is illustrating is the proactive application of the tools, methods, and best practices to prevent problems during marketing work.

Traditional tools, methods and best practices from DMAIC and Design for Six Sigma we will adaptively use:

1. Project planning, scoping, cycle-time design, and management methods
2. Voice of the Customer gathering and processing methods
   a. Customer identification charts, customer interview guides, and customer interviewing techniques
   b. Jiro Kawakita (KJ) Voice of the Customer (VOC) structuring, ranking, and validation methods
3. Competitive benchmarking studies
4. Customer value management tools
   a. Market-perceived quality profiling
   b. Key events timelines
   c. What-who matrix
   d. Won-Lost Analysis

5. Strength-Weakness-Opportunity-Threat (SWOT) Analysis
6. Porter's 5 Forces Analysis
7. Real-Win-Worth Analysis
8. Survey and questionnaire design methods
9. Product portfolio architecting methods
10. Requirements translation, ranking, and structuring
    a. Quality Function Deployment (QFD) and the houses of quality
11. Concept generation methods
    a. Brainstorming, mind mapping, and TRIZ
12. Pugh Concept Evaluation & Selection Process
13. Critical Parameter Management
14. Process maps, value chain diagrams, customer behavioral dynamics maps, work flow diagrams, critical path cycle-time models
15. Cause & Effect Matrices
16. Failure Modes & Effects Analysis
17. Noise diagramming (sources of variation)
18. Basic and inferential statistical data analysis
    a. Normality Tests, graphical data mining (data distribution characterization), hypothesis testing, t-tests, confidence intervals, sample sizing; ANalysis Of the VAriance (ANOVA), and regression
19. Multivariate Statistical Data Analysis
    a. Factor, discriminate, and cluster analysis
20. Measurement System Design & Analysis
21. Statistical Process Control

22. Capability studies
23. Design of transfer functions: analytical math models ($Y = f(X)$ from first principles, as well as empirical data sets from designed experiments)
24. Monte Carlo Simulations ($\Delta Y = f(\Delta X)$)
25. Sequential Design of Experiments (including multi-vari studies, full and fractional factorial designs for screening and modeling, conjoint studies, robust design Design of Experiments (DOE) structures)
26. Control plans

As you can see with at least 26 tools, methods, and best practices to creatively adapt and apply to marketing tasks within the flow of marketing work across an enterprise, there is a huge opportunity to prevent problems and achieve growth goals as we selectively design our marketing work.

Acknowledgments

Thanks to our friends and colleagues who took the time out of their hectic schedules to read early manuscript drafts and provide insights and improvement suggestions. In particular, we would like to thanks John S. Branagan, Marianne Doktor, Cecelia Henderson, John Loncz, and Deborah Pearce.

We would like to thank all the great people at Prentice Hall for their support and hard work to make this part of the Six Sigma for Growth series a success: Suzette Ciancio, marketing; Heather Fox, publicist; Bernard Goodwin, editor; Christy Hackerd, project editor; Michelle Housley, editorial assistant; Gayle Johnson, copy editor; and Marty Rabinowitz, production.

Burke also would like to acknowledge Donald Carli and Paul McCarthy.

About the Authors

Clyde M. Creveling is currently president of Product Development Systems & Solutions, a full-service product development consulting firm. Prior to that he was an independent consultant, DFSS product manager, and DFSS project manager with Sigma Breakthrough Technologies, Inc.

Mr. Creveling was employed by Eastman Kodak for 17 years as a product development engineer in the Office Imaging Division. He also spent 1 1/2 years as a systems engineer for Heidelberg Digital as a member of the System Engineering Group. During this 18-year period, he worked in R&D, product development/design/ system engineering, and manufacturing. Mr. Creveling has five U.S. patents.

He was an assistant professor at Rochester Institute of Technology for four years, developing and teaching undergraduate and graduate courses in mechanical engineering design, product and production system development, concept design, robust design, and tolerance design. Mr. Creveling is a certified expert in Taguchi Methods.

He has lectured, conducted training, and consulted on product development process improvement, product portfolio definition and development, design and marketing for Six Sigma methods, technology development for Six Sigma, Critical Parameter Management, robust design and tolerance design theory and applications in numerous U.S., European, and Asian locations. His clients include 3M, Merck & Co., Motorola, Samsung, Applied BioSystems, United Technologies, ACIST Medical Systems, Beckton Dickenson, Mine Safety Appliances, Callaway Golf, Lightstream, Kodak, NASA, Iomega, Xerox, Sequa Corp. (Atlantic Research Corp., MEGTEK, Sequa Can), GE Medical Imaging Systems, Bausch & Lomb, Moore Research, IIMAK, Case—New Holland, Maytag, Cummins, Schick, Purolator, Goulds Pumps, INVENSYS, Shaw Carpet, Heidelberg Digital, Nexpress, StorageTek, and Universal Instruments Corporation. He has been a guest lecturer at MIT, where he assisted in the start-up of a graduate course in robust design within the MS in System Development and Management program.

Mr. Creveling coauthored *Engineering Methods for Robust Product Design* (Addison-Wesley, 1995, ISBN 0-201-63367-1) and wrote the world's first comprehensive text on developing tolerances, *Tolerance Design* (Addison-Wesley, 1997, ISBN 0-201-63473-2). This book focuses on analytical and experimental methods for developing tolerances for products and manufacturing processes. Mr. Creveling also is the coauthor of a new text on Design for Six Sigma for Prentice Hall's Engineering Process Improvement series, *Design for Six Sigma in Technology and Product Development* (ISBN 0-130-09223-1).

Lynne Hambleton is a business consultant providing advisory and business implementation services, with special focus on operational process improvement and growth initiatives, commercialization, change management, interim general management, marketing, and business plan development. She has held management positions in a Fortune 100 company, the public sector, and start-ups, gaining experience in general management, marketing, operations, strategic planning, alliance development, and sales/channel management. She also has served as an adjunct professor of strategic planning at Rochester Institute of Technology's School of Business.

Ms. Hambleton received a master's degree in business administration with an emphasis in industrial marketing, a master's degree in adult and higher education/organizational development, and a bachelor of science degree in psychophysiology, all from the University of North Carolina—Chapel Hill. Ms. Hambleton has been a PMI-certified Project Management Professional (PMP) since 1998.

Ms. Hambleton's additional publications include the chapter "Supporting a Metamorphosis Through Communities of Practice" in *Leading Knowledge Management and Learning* by Dede Bonner (2000) and the article "How does a company the size of Xerox design a curriculum in project management for the entire organization?" in *In Search of Excellence in Project Management*, Volume 2 by Harold Kerzner (1999).

Ms. Hambleton lives in Rochester, NY with her husband, Bill, and their two sons, Corbin and Garrett.

Burke McCarthy has marketed industrial and consumer products in a wide range of industries, including photography, digital imaging, printing, telecommunications, maritime transportation, pharmaceuticals, soaps, fragrances, medical and diagnostic equipment, and HVAC.

Mr. McCarthy earned an MBA in finance from Seton Hall University in 1988. His career with Eastman Kodak took him from technical sales representative in New York City to product line manager in Rochester, NY to regional manager and vice president in Los Angeles to director, global strategic growth to marketing manager of Kodak single-use camera products in Rochester.

His other roles and responsibilities have included technical sales; product development; director of marketing; president, business development; sales management; and strategic planning. He has also worked at Foveon of Santa Clara, Calif. and at Xerox as strategic accounts manager in 2000 based in Rochester. In 2003, he assumed the role of vice president, Six Sigma for Marketing and Sales at PDSS, Inc.

1

Introduction to Six Sigma for Marketing Processes

Marketing in Product Portfolio Renewal, Commercialization, and Post-Launch Product Line Management

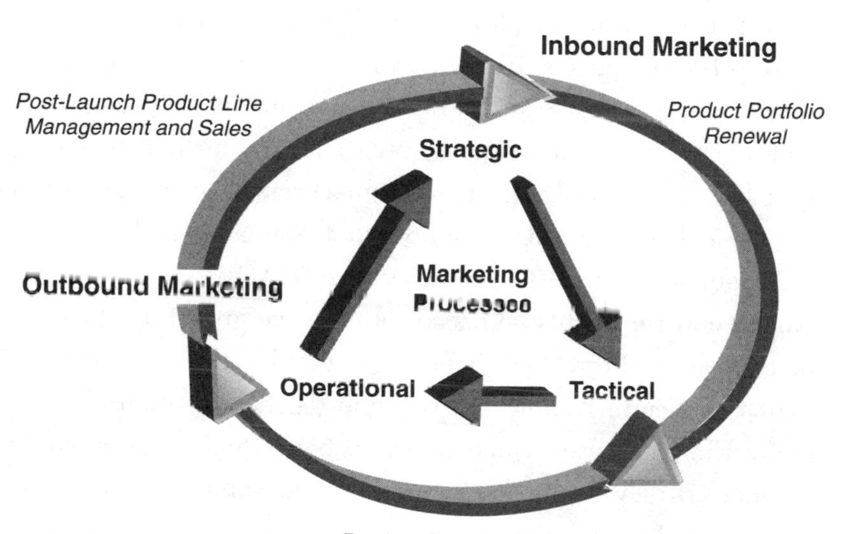

Growth and Innovation

Imagine the possibilities if you possessed a crystal ball that let you predict the future. You would know what will work and what won't work to create and sustain growth. You would know when to correct for competitive and environmental changes and how to prevent going off-course. Is this a fantasy? Can a business predict (with some certainty) what will drive success and how to stay on the right track? We believe the answer is yes. The appropriate data can inform executives, with high probability, whether the critical elements of the business are performing as planned to achieve desired results.

Performance against plan is how a business typically defines success. Businesses gauge success by a multitude of metrics—revenue, income, profit, customer satisfaction, market share, return on equity, return on assets, return on investments, and so on. Bottom-line, planned success means reaching and sustaining goals over time—usually growth goals. The challenge lies in determining the vital few results to focus on and the critical metrics that best monitor performance. The Fortune 500 list serves as another metric of success. Of the top 100 companies, 70 have been in the top 100 for five or more years. Interestingly, 63% of those 70 companies acknowledge implementing Six Sigma to some degree. Through further analysis, we have found that these same 44 Six Sigma users also reported on average 49% higher profits (compounded annually) and 2% higher Compounded Annual Growth Revenue (CAGR) than their peers. Notice how the profits outpaced the revenue growth for this group of companies. More than likely, they employ the "traditional" Six Sigma cost-cutting approach. Imagine the benefit these firms will enjoy when they also begin to apply Six Sigma to the top line to drive revenue. If they deploy Six Sigma into marketing and sales with as much discipline and rigor as they did to eliminate waste in manufacturing and engineering, these firms' CAGR will outrun their competitors as much as their profits have, and they will easily secure a prominent spot on the Top 100 list for another five or more years.

Benchmarking tells us that successful companies, which effectively implement Six Sigma tools, methods, and best practices find the following benefits:

- **Systematic innovation:** Generate and define more ideas linked with market opportunities in a structured way.
- **Manage risk better:** Identify critical issues early in the commercialization process such that plans can be developed to mitigate or eliminate risk going forward.
- **Higher return yield from a project portfolio:** Avoid overloading resources with too many low-risk, small-gain projects through a discriminating selection process. Select fewer projects—the "best fit" projects, not necessarily the easiest projects.

Business leaders often hold marketing and sales accountable for driving revenue growth—the panacea for most business ills. They want these teams to improve their accuracy rate of committing to, and achieving, their goals. Marketing executives seek new ideas to bolster their success rate. Applying Six Sigma to marketing may be a new approach, but it comes with an "insurance policy." Six Sigma has a proven track record in other parts of the business. Six Sigma concepts can provide additive elements to increase the competitive advantage marketing needs to act proactively, sustain its positive momentum, and keep pace with the ever-changing landscape.

To tailor Six Sigma to marketing, you start with an overview of how it works. We find that marketing professionals rarely view their own work as process-oriented; it often is depicted as project- or activity-based. However, the American Marketing Association (AMA) defines "marketing" as "a set of *processes* for *creating*, *communicating*, and *delivering* value to customers and . . . managing customer relationships in ways that benefit the organization and . . . stakeholders." The *American Heritage Dictionary* describes a "process" as a "*series* of *actions*, *changes*, or *functions* bringing about a result" and a "function" as "something closely related to another thing and dependent on it for its existence, value, or significance." Others define "marketing" as the *process* to identify, anticipate, and then meet customers' needs and requirements. This definition seems narrow. In a special issue of *Journal of Marketing* (1999, Volume 63, pp. 180–197), Christine Moorman and Roland Rust propose that

> **the marketing function should play a key role in managing several important connections between the customer and critical firm elements, including connecting the customer to (1) the product, (2) service delivery, and (3) financial accountability. . . . Marketing's value . . . is found to be a function of the degree to which it develops knowledge and skills in connecting the customer to the product and to financial accountability.**

Hence, to fully capture marketing's value, the customization of Six Sigma should span the scope of connecting the customer to the product and to financial accountability.

Moorman and Rust's research suggests that the value of the marketing function is due to *how well-developed the methodologies are for facilitating the customer-product connection*. Marketing's customer-financial accountability linkage often is not well understood, but it needs to account for *profitability considerations in attracting and retaining customers*. It is not about cost; it is about profitable growth. Ideally, marketing should effectively and efficiently create and sustain growth for the firm. How is that best done? A challenge is to determine which marketing methodology best facilitates the customer-product-financial linkages. The marketing methodology should nurture and channel the firm's important creativity and growth capabilities.

The Six Sigma discipline gives business leaders the opportunity to drive more fact-based decisions into managing the business. Six Sigma has been successfully applied to the technical aspects of a business (such as engineering and manufacturing). A new effort is afoot to bring Six Sigma into the "softer" side of business—marketing. By adding more "science" to the "art" of marketing, the Six Sigma approach can be the next best thing to a crystal ball.

A decision-making process that lacks the appropriate facts causes leaders to fill the void with *intuition*. If facts are absent, statistically grounded probabilities can strengthen decision-making. Marketing executives should shed their use of intuition (or "gut feeling") to solve business issues and/or drive growth. Columnist and author Marilyn Savant said, "Not knowing the difference between opinion and fact makes it difficult to make decisions. . . ." Intuition sneaks into every business at some point. The objective is to recognize it when it appears and to deal with it directly by using facts to support or deny the "hypothesis." Bernard Baruch, an advisor to six U.S. presidents, said, "Every man has the right to be wrong in his opinions. But no man has a right to be wrong about his facts. . . ."

The Six Sigma concept has evolved over the past several decades to represent a set of fundamental business concepts that puts customers first and uses fact-based decision-making to drive improvements. It was first used in the U.S. at Motorola to cut costs by reducing variation in manufacturing. This book represents the next evolution of Six Sigma—a marketing application. We believe a unique view of Six Sigma's techniques and tools can be applied to drive income growth. It is our experience that companies are only beginning to implement Six Sigma to drive sales and marketing; however, the IDEA is increasingly discussed. In the fall of 2005, the Worldwide Conventions and Business Forums (WCBF) held its second annual conference on Six Sigma in sales and marketing. This is a cutting-edge application of Six Sigma.

This book focuses on the new frontier of applying the Six Sigma discipline to an integrated, enterprise-wide strategy to create measurable capabilities in sustaining top-line growth. This book can be read on two different levels. First, it introduces marketing managers and executives to Six Sigma (at a high level) and suggests a unique approach to applying its concepts to marketing. Second, for those familiar with Six Sigma, this book suggests a unique, flexible combination of tools and techniques tailored for marketing. Regardless of which audience you may find yourself in, we trust that this book contains new thinking and practical recommendations that will yield success.

Six Sigma has been successfully applied to engineering and manufacturing. Adding more "science" to the "art" of marketing offers a number of benefits, including project selections aligned with attractive market opportunities, a faster and more accurate product commercialization process, and better cross-functional communication. The Six Sigma approach of using proven tools, methods, and best practices across the entire marketing process can be the next best thing to a crystal ball because, with time and experience, it can deliver more predictable outcomes.

What Is Six Sigma?

The term "Six Sigma" has several meanings. At the most encompassing level, a corporation can define it as its philosophy—a way of thinking. By doing so, a company's management structure, employee roles, and operations are defined, in part, by this fact-based discipline. Or it can be defined as a method and tool set—for example, using the Define-Measure-Analyze-Improve-Control (DMAIC) technique to make improvements and solve problems *within an existing process*. Or, at the simplest level, it can be defined as a specific statistical quantity, describing the number of defects produced due to variation in a product or process. Technically, Six Sigma is described as a data-driven approach to reduce defects in a process or cut costs in a process or product, as measured by "six standard deviations" between the mean and the nearest specification limit. "Sigma" (or σ) is the Greek letter used to describe variability, or standard deviation, such as defects per unit. Figure 1.1 shows a normal distribution of a

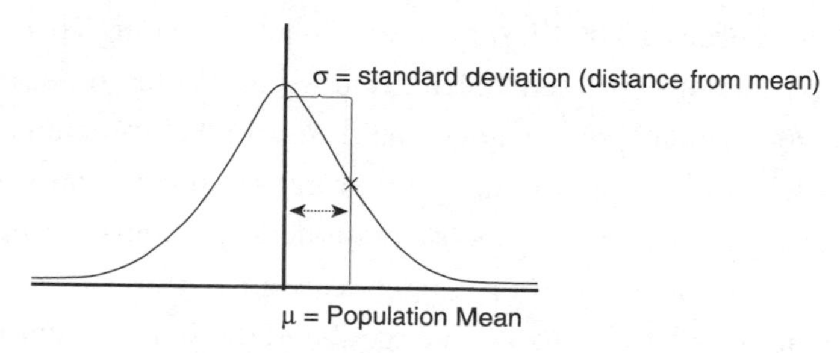

FIGURE 1.1 A normal distribution.

population, with its mean (μ) in the center and a data point on the curve indicating one standard deviation (1σ) to the right of the mean.

How well a desired outcome (or target) has been reached can be described by its mathematical *average*; however, this may be misleading. The average of a data set masks the variation from one data point to the next. The *standard deviation* describes how much variation actually exists within a data set. An average is mathematically defined as the sum of all the data points divided by the number of data points. This is also called an *arithmetic mean*. The *standard deviation* is calculated as the square root of the variance from the mean.

Why is the number six frequently coupled with the word "sigma"? If a process is described as within "six sigma," the term quantitatively means that the process produces fewer than 3.4 defects per million units (or opportunities). That represents an error rate of 0.0003%; conversely, that is a defect-free rate of 99.9997%. That's pretty good, right? Professional marketers can relate to this because they see errors and can exploit the opportunity to reduce variation and its effects on results.

What level of variance (or error rate) in a process should you accept? If the resulting process data is within three standard deviations (3σ) from the mean, is that good or bad? The answer depends on your business. Let's say you are in the shipping business, and you experience only a 1% error rate for every million deliveries. Is that good? That translates into a 99% error-free business (or a four-sigma level [4σ]), or 6,210 defects per million. Is that good? In business terms, that means 20,000 lost pieces of mail per hour. That could cause some serious customer satisfaction issues. Within other industries, a “four-sigma” performance could mean 6,800 problems with airplane takeoffs per month, or 4,300 problems in common surgical procedures per week, or no electricity for almost 7 hours per month. Remember, the sigma measure compares your performance to customer requirements (defined as a target), and the requirement varies with the type of industry or business.

That is a brief technical description of Six Sigma. The concepts put forth in this book (and the literature) go beyond a mathematical discussion and extend into how companies deploy these statistical tools—as a business initiative. Successfully implementing the Six Sigma approach requires companies to consider changes in methodologies across the enterprise, introducing new linkages. Similar to the Total Quality Management (TQM) initiative, some benchmark companies create new employee roles (such as Black Belt project leaders). Some also institute a new management or organizational structure and new or revised project and operational processes to instill the concept.

Three benchmark examples of how Six Sigma permeates a corporate philosophy and becomes a business initiative can be found by studying Motorola, Allied Signal, and General Electric (GE). Motorola created Six Sigma (largely attributed to Bill Smith) as a rallying point to change the corporate culture to better compete in the Asia-Pacific telecommunications market. At that time, Motorola’s main focus was on manufacturing defect reduction. Allied Signal rebuilt its business with bottom-line cost improvement using Six Sigma. Eventually Allied extended its Six Sigma implementation into its business and transactional processes for cost control. GE revolutionized how an entire enterprise disciplines itself across its operations, transactions, customer relations, and product development initiatives. GE implemented Six Sigma at the Customer for the customer and top-line growth using an approach called Design for Six Sigma, a methodology for product creation and development.

These three benchmark companies are pioneers in the traditional application of Six Sigma. They adhered to the three Six Sigma fundamentals of tool-task linkage, project structure, and, most importantly, result metrics. Before we explore the new growth-oriented Six Sigma for marketing, let's review Six Sigma's original methods (see Figure 1.2). This background information will help you understand how practitioners repair an inefficient or broken marketing process.

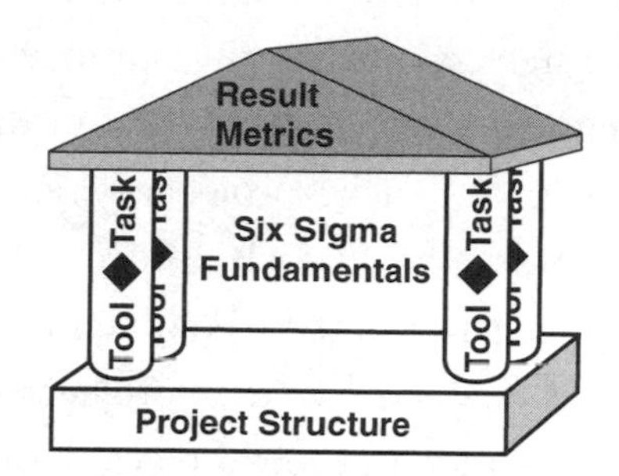

FIGURE 1.2 Six Sigma fundamentals.

The Traditional Six Sigma Approach

The Six Sigma concept started out as a problem-solving process. The problems generally concerned eliminating variability, defects, and waste in a product or process, all of which undermine customer satisfaction. Six Sigma practitioners call this original method DMAIC (pronounced "duh-may-ick")—Design, Measure, Analyze, Improve, and Control. The five steps are as follows:

1. **Define** the problem.
2. **Measure** the process and gather the data that is associated with the problem.
3. **Analyze** the data to identify a cause-and-effect relationship between key variables.
4. **Improve** the process so that the problem is eliminated and the measured results meet existing customer requirements.

5. **Control** the process so that the problem does not return. If it does return, it should be controllable using a well-designed control plan.

The DMAIC process is easy to learn and apply. It provides strong benefits to those who follow its simple steps using a small, focused set of tools, methods, and best practices. The original pioneer of Six Sigma, Motorola, used the approach to eliminate variability in its manufacturing process and better meet basic market requirements. Companies that find success in using this approach train small teams to adhere to this approach without wavering in their completion of specific project objectives. These projects typically last six to nine months. Companies learn the DMAIC process and apply the tools much like a well-trained surgical team conducting an operation. They are focused, they are enabled by their project sponsors, and they deliver on the goals specified in their project charter.

The key elements in a DMAIC project are team discipline, structured use of metrics and tools, and execution of a well-designed project plan that has clear goals and objectives. When large numbers of people across a multinational company use the simple steps of DMAIC, objectives and result targets are much harder to miss. If everyone solves problems differently, nonsystematically, they become one-offs. Company-wide process improvement initiatives break down. Cost and waste reduction are usually haphazard. The corporation has difficulty integrating and leveraging the improvements across the enterprise. In this undisciplined environment, cost reduction and control are unpredictable and unsustainable.

Lean Six Sigma modifies the DMAIC approach by emphasizing speed. Lean focuses on streamlining a process by identifying and removing non-value-added steps. MIT pioneered the Lean approach in a manufacturing environment. A "leaned production" process eliminates waste. Target metrics include zero wait time, zero inventory, scheduling using customer pull (rather than push), cutting batch sizes to improve flow, line balancing, and reducing overall process time. Lean Sigma's goal is to produce quality products that meet customer requirements as efficiently and effectively as possible. This can be readily applied to the process steps to develop sales collateral or participation in a trade show.

If a process cannot be improved as it is currently designed, another well-known Six Sigma problem-solving approach can be applied. The DMADV process is used to fundamentally redesign a process. Sometimes it may also be used to design a new process or product when new requirements emerge. The five steps are as follows:

1. **Define** the problem and/or new requirements.
2. **Measure** the process and gather the data that is associated with the problem or in comparison to the new requirements.
3. **Analyze** the data to identify a cause-and-effect relationship between key variables.
4. **Design** a new process so that the problem is eliminated or new requirements are met.
5. **Validate** the new process to be capable of meeting the new process requirements.

A second redesign approach has been developed to incorporate elements from a Lean Six Sigma approach—the **DMEDI** process. This methodology is essentially similar to DMADV, but it uses a slightly different vocabulary and adds tools from the Lean methodology to ensure efficiency or speed. The steps are as follows:

1. **Define** the problem or new requirements.
2. **Measure** the process and gather the data that is associated with the problem or new requirements.
3. **Explore** the data to identify a cause-and-effect relationship between key variables.
4. **Develop** a new process so that the problem is eliminated and the measured results meet the new requirements.
5. **Implement** the new process under a control plan.

Whether you use DMADV or DMEDI, the goal is to design a new process to replace the incapable existing process. This is still the classic Six Sigma for problem-solving. The classic methods aim to improve processes and get them under control. They all build on similar fundamentals:

- Tool-task linkage
- Project structure
- Result metrics

Once this is done, however, another form of a Six Sigma-enabled process is required to expand beyond problem-solving.

The new frontier for Six Sigma is in *problem prevention*, which should occur as part of your daily workflow. As they say, an ounce of prevention is worth a pound of cure. Six Sigma for Marketing and Six Sigma for Sales, like Design for Six Sigma and Six Sigma for Research and Technology Development, are structured tools-tasks-deliverables sets for problem prevention during the phases and gates of product portfolio definition and development, research and technology development, product commercialization, and post-launch product-line management processes.

The traditional "reactive" DMAIC and Lean methods should be used for their intended purposes—to reduce variances, cut costs, and streamline processes. We mean no disrespect when using the terms "traditional" or "old-style." We are trying to define the future of Six Sigma. By necessity, we have to draw a distinction between the original application and a new approach that transcends problem-solving, cost-cutting, and reactive methods. The emerging application of Six Sigma builds on the fundamentals but travels on a different financial journey—seeking top-line growth. Controlling costs is important, but creating sustainable growth is equally important, if not more so. When all you have is a hammer, everything looks like a nail. Use the appropriate tool for a given task. Both the traditional and new Six Sigma methods add value. *Use the right tool, at the right time, to help ask and answer the right questions.*

Applying Six Sigma to Marketing

Marketing professionals want to avoid suppressing creativity with tools and structure. Process-centric work may at first seem slow, routine, and burdensome. Moreover, marketing may think statistical analysis can dampen spontaneity and innovation. But our experience suggests that the opposite is true. The Six Sigma model described in this book plans for innovation and creativity to occur. If implemented correctly, a proven methodology averts rework (caused by mistakes), ensures completeness, and reinforces quality standards. A well-constructed method that requires improvement should plan for innovation and identify the appropriate participants. Moreover, Six Sigma can help tackle the new, the unique, and the difficult.

Few dispute the value of measurement. However, that which is easily measured rarely produces real or optimal value. Real value comes from measuring what others cannot or will not measure. This brings to mind a lesson from history. In 1726, Benjamin Franklin wondered if that warm swath of water he noticed crossing the North Atlantic had anything to do with the longer times it took to sail from England to the U.S. Franklin's cousin, Tim Folger, a whaler, knew that sailing around the current as if it were a mountain was much faster than sailing directly through the current to Philadelphia. In 1769, Franklin sold charts in London on "how to avoid the Gulph [sic] Stream" that cut westbound travel time up to 50%. To this day, Folger's map is surprisingly accurate. These measures gave Folger's whaling business a competitive advantage and higher revenue margins.

The benefit of integrating Six Sigma into your marketing processes includes better information (management by fact) to make better decisions. Using the more robust approach reduces the uncertainty inherent in marketing—a creative, dynamic discipline. Go-to-market processes with Six Sigma embedded in them can better sustain growth. One way to maintain growth over time is to focus on "leading" indicators of your desired goal. Leading indicators are factors that precede the occurrence of a desired result. Let's say you are concerned about dealing with a weight-induced disease such as a heart attack or diabetes. You could be reactive by regularly getting on the scale to see how much you weigh. Or you could be proactive by monitoring your caloric intake and burn rate. The latter approach of watching what you eat and how much energy you expend during exercise is harder than simply getting on the scale. The latter approach monitors "leading" indicators—critical activities that occur *before* weight gain. The "lagging" indicator takes a snapshot after the occurrence of an event. Lagging indicators force you into a reactive response if the results fail to meet the target. The act of losing weight may be more difficult than measuring the leading indicators of caloric intake and burn rate. The advice of "pay me now or pay me later" comes to mind.

Business lagging indicators involve measuring defects, failures, and time. Lagging indicators can include functional performance measures such as Unit Manufacturing Cost (UMC), quality measures such as Defects Per Million Opportunities (DPMO), and time-based measures of reliability such as Mean Time Between Failures (MTBF). Lagging indicators for marketing include market share and revenue—common performance metrics. A powerful leading indicator is customer satisfaction *before* a sales transaction (such as satisfaction with an information meeting or advertising piece). Another leading indicator may be the distribution channel's satisfaction with a product (or samples), whereby the salespeople want to use it themselves. Leading indicators help you anticipate whether you will hit the target. Since leading indicators occur *before* the desired result, you can be proactive in "correcting" poor performance. Armed with this knowledge, marketing can examine initiatives from a different perspective. To drive and sustain growth, performance and quality metrics need to be proactive rather than reactive. (Examples of continuous data include cycle time, profit, mass, and rank [customer satisfaction scores on a scale of 1 to 10]. Continuous variables are more informative and describe a process better than discrete or attribute data. Examples of discrete or attribute data include binary [yes/no, pass/fail] and counts [the number of defects].) Leading-indicator data, when established as a continuous variable, requires far fewer data samples to draw conclusions and make a decision as opposed to discrete-failure data.

Recall that a marketing methodology should facilitate the customer-product-financial linkages. This requirement seeks a comprehensive scope of marketing's responsibilities from offering inception, through offering development, to the customer experience. This comprehensive scope encompasses a business's strategic, tactical, and operational aspects. Marketing's role in each of these three business areas can be defined by the work it performs in each. This work can be characterized by a process unique to each. These three processes define how marketing's work links the strategic, tactical, and operational areas in a closed-loop fashion, as shown in Figure 1.3.

FIGURE 1.3 The strategic-tactical-operational triangle.

Let's examine the process that resides in each area. The *Strategic Planning and Portfolio Renewal* process defines a business's set of marketplace offerings. This strategic activity is fundamental for an enterprise, because it refreshes its offerings to sustain its existence over time. Multiple functional disciplines may be involved in this process, or the enterprise may limit this work to a small set of corporate officers, depending on the size of the enterprise and the scope of its offerings. This process generally calls for a cross-functional team composed of finance, strategic planning, and marketing, and sometimes research, engineering, sales, service, and customer support. A business with a unique strategic planning department may use it as a surrogate for the other various functional areas. If this is the case, the strategy office typically includes people with various backgrounds (research, finance, and marketing). This process can span a year and should get updated on a regular basis. Portfolio planning and management are the foundation from which to build and grow a business. Our experience tells us that successful businesses have marketing play a key role in the Strategic Planning and Portfolio Renewal process. In his book *Winning at New Products*, Robert G. Cooper states

> **There are two ways to win at new products. One is to do projects right—building in Voice of the Customer, doing the necessary up-front homework, using cross-functional teams . . . The other way is by doing the right projects—namely, astute project selection and portfolio management.**

Six Sigma can help improve performance in this area.

The *Product* and/or *Services Commercialization* process defines the tactical aspects of a business. This process defines, develops, and readies a business's offering for the marketplace. The industry, market segment, and size/scale/complexity of the offering dictate the number of functional disciplines involved in this process and the amount of time it spans. The time frame ranges from several months to several years. A business usually manages this process by establishing a unique *project* team to develop a single product or services from the portfolio of opportunities. At a minimum, two types of disciplines are needed—technical functions to drive content and customer-facing functions. The technical experts develop the offering and may include engineering, research, and manufacturing. The customer-facing disciplines represent roles along the value chain that interface with a business's customer or client, such as marketing, sales, services, and customer support. In the Commercialization process, marketing may represent the customer-facing touch points throughout the process and may bring in the other functional areas toward the conclusion of the process in preparation for handoff to ongoing operations.

The *Post-Launch Operational Management* process unifies the operational aspects of a business across the value chain. This process represents long time frames (often years), depending on the life cycle of a given offering (product or service). The offering and go-to-market strategy dictate the variety of functional disciplines involved across the value chain. Again, marketing may play a representative role, integrating multiple functional areas as it manages the product line (or offering) throughout its life cycle.

Marketing professionals typically view their function as a *set of activities* or *projects* rather than a set of processes. It may seem unnatural at first to think about marketing work in terms of a process. However, process thinking provides an easily communicated road map that can describe interactivity with other processes. For example, marketing's tactical Product Commercialization process can cleanly map to the technical community's Product Design and Development process. By creating this linkage, the two functions better understand their interdependency with one another and can speak a common language as the output of one process becomes the input of the other's process. This book is a guide for leaders in the design of Six Sigma-enabled marketing processes.

The book *The Innovator's Solution*, by C. Christensen and M. Raynor, addresses the importance of process thinking. Similar to a business executive forecasting next quarter's performance, the authors ask the reader to predict the next two numbers in two different sequences. The first sequence of numbers is 3, 5, 7, 11, 13, 17, ___, ___. The second sequence of numbers is 75, 28, 41, 26, 38, 64, ___, ___. Do you know the answers? Without knowing the process that describes the sequence, you can only guess with little or no certainty. The answers for the first sequence are 42 and 6. This sequence was determined by tumbling balls in a drum being selected for an eight-number lottery winning. The answers for the second sequence are 2 and 122. They were determined by the sequence of state and county roads found along a scenic route in northern Michigan, heading toward Wisconsin. Christensen and Raynor point out that "results alone cannot predict future outcomes. The process itself must be understood to predict outcomes." Imagine the increased value that marketing could provide if it could improve its ability to predict the results of its work.

To recap, process thinking is used throughout this book. We explore applying Six Sigma concepts to the work of marketing. Marketing professionals' work environment on a day-to-day basis is *not* a DMAIC-based workflow structure. Marketing's work breaks down into the fundamental process of three key business arenas:

- **Strategic area:** The Portfolio Renewal process.
- **Tactical area:** The Commercialization process (commercializing a specific product and/or service).
- **Operational area:** The Post-Launch Line Management process (managing the launched portfolio) and its go-to-market resources throughout its life cycle, across the value chain.

The natural flow of marketing work starts with strategic renewal of the offering portfolios, to the tactical work of commercializing new offerings, and finally to the operational work of managing the product and services lines in the post-launch sales, support, and service environment. Marketing professionals frequently overlook the fact that their contributions are part of a process (or a set of related processes). They view their work as part of a program or project. However, marketing work can be repeated. The time frame for repetitiveness may extend over a year or more, but nonetheless, the work is procedural in nature. (The American Society for Quality [ASQ] defines a process as "a set of interrelated work activities characterized by a set of specific inputs and value-added tasks that make up a procedure for a set of specific outputs.") Most marketers would agree that "strategic planning" and "launching a product" meet this "process" definition. The Six Sigma approach embraces a process view to communicate its structure and flow of interrelated tasks. Although it may seem unnatural to marketing professionals, the best way to describe Six Sigma for Growth is through a process lens.

The strategic and tactical areas are internally focused; hence, we refer to them as *inbound* marketing areas. External data is critical to successful portfolio definition and development, and product commercialization. However, the output of those processes is intended for internal use. These process outputs are not yet ready for external consumption. The outputs that are ready for prime-time market exposure are part of *outbound* marketing. The operational processes involving post-launch product marketing, sales, services, and support are customer-facing activities. Given the different customers of inbound and outbound marketing, the requirements for each differ. These requirements ultimately define the success (or failure) of the deliverables.

Problems can be prevented in inbound as well as outbound marketing processes. Inbound marketing focuses on strategic product portfolio definition and development, and tactical product commercialization. Inbound marketing can cause problems by underdeveloping the right data needed to renew product portfolios. The data is needed to define specific new product requirements, thereby directing commercialization activities. And inbound marketing data defines launch plans, which determine downstream operational success. You can design and launch the wrong mix of products and hence miss the growth numbers promised in the business cases that were supposed to support the company's long-term financial targets.

Outbound marketing is focused on customer-facing operations. It encompasses post-launch product line management across the value chain (sales and services, including customer support). Outbound marketing can create problems and waste by failing to develop the right data to make key decisions about managing, adapting, and discontinuing the various elements of the existing product and service lines. Outbound marketing also could fail to get the right information back upstream to the product portfolio renewal teams. They need to renew the portfolio based on real, up-to-date data and lessons learned from customer feedback and the marketing and sales experts in the field.

The importance of the comprehensive, closed-loop strategic-tactical-operational scope provided the structural underpinnings used to create the unique Six Sigma methods for marketing. Each of these arenas has a flow of repeatable work—a process context that is quite different from the steps found in the traditional Six Sigma methods. However, the fundamental Six Sigma elements from the classic approaches have been maintained: tool-task linkage, project structure, and result metrics. This new work is made up of specific tasks that are enabled by flexible, designable sets of tools, methods, and best practices. The strategic, tactical, and operational processes within an enterprise align with phases that can be designed to prevent problems—to limit the accrual of risk and enable the right kind and amount of data to help make key decisions. The traditional methods help you improve and redesign your processes and get them under control. If the objective is to renew portfolios, commercialize products, or manage product lines, a different approach is required that employs a different set of steps we call *phases*.

Unique Six Sigma Marketing Methods

A unique Six Sigma marketing method was created for each of the three areas: strategic, tactical, and operational. The method to guide marketing's strategic work is called IDEA. The approach for tactical work is called UAPL. The method to direct marketing's operational work is called LMAD. Each method has a chapter devoted to it, detailing its unique combination of tools-tasks-deliverables.

The strategic marketing process environment has the following four distinct phases, known as the IDEA process for portfolio renewal and refresh:

1. **Identify** markets, their segments, and the opportunities they offer.
2. **Define** portfolio requirements and product portfolio architectural alternatives.

3. **Evaluate** portfolio alternatives against competitive portfolios by offering.
4. **Activate** ranked and resourced individual commercialization projects.

The tactical marketing process environment has the following four distinct phases, defined as the UAPL process for specific product and/or service commercialization projects:

1. **Understand** the market opportunity and specific customer requirements translated into product (or service) requirements.
2. **Analyze** customer preferences against the value proposition.
3. **Plan** the linkage between the value chain process details (including marketing and sales) to successfully communicate and launch the product (or service) concept as defined in a maturing business case.
4. **Launch:** Prepare the new product (or service) under a rigorously defined launch control plan.

The operational marketing process environment has the following four distinct phases. This process is called the LMAD process for managing the portfolio of launched products and/or services across the value chain:

1. **Launch** the offering through its introductory period into the market according to the launch control plan of the prior process.
2. **Manage** the offering in the steady-state marketing and sales processes.
3. **Adapt** the marketing and sales tasks and tools as "noises" require change.
4. **Discontinue** the offering with discipline to sustain brand loyalty.

Each of these processes features distinct phases in which sets of tasks are completed. Each task can be enabled by one or more tools, methods, or best practices that give high confidence that the marketing team will develop the right data to meet the task requirements for each phase of work. A *Gate Review* at the end of a phase is commonly used to assess the results and define potential risks (see Figure 1.4). Marketing executives and professionals find phase-gate reviews an important part of risk management and decision-making. In the post-launch environment, gates are replaced by *key milestone reviews* because you are in an ongoing process arena—unlike portfolio renewal or commercialization processes, which have a strictly defined end date.

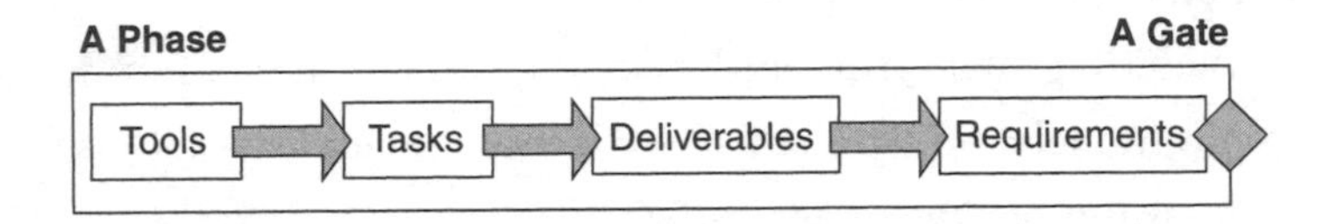

FIGURE 1.4 The tools-tasks-deliverables-requirements linkage.

This book describes how Six Sigma works in the context of strategic, tactical, and operational marketing processes. It focuses on integrating marketing process structure, requirements, and deliverables (phases and gates for risk management), project management (for design and control of marketing task cycle time), and balanced sets of marketing tools, methods, and best practices.

Recall that if a marketing process is broken, incapable, or out of control, you should use one of the traditional Six Sigma approaches to improve or redesign it. This book assumes that the strategic, tactical, and operational marketing processes have been designed to function properly. This book answers the question of *what to do* and *when to do it* within structured marketing processes.

Marketing processes and their deliverables must be designed for efficiency, stability, and, most importantly, measurable results—hence the importance of Six Sigma. We will work within the IDEA, UAPL, and LMAD processes, applying their accompanying tool-task sets to create measurable deliverables that fulfill the gate requirements. You may choose to call your process phases by different names—that's fine. *What you do and what you measure* are what really matter.

Throughout this book, the word "product" refers to a generic company "offering" and represents a *tangible product* and a *services offering*. This book discusses technology-based products frequently, because of marketing's interdependency with the technical community. In parallel, R&D, design, and production/services support engineering should use growth- and problem-prevention-oriented forms of Six Sigma in their phases and gates processes. The Six Sigma approach serves as a common language between the marketing and technical disciplines. The term "solutions" usually involves both technology and services; thus, "product" and "service" encompass the scope of a given solution. Regardless of the offering, the Six Sigma approach we are outlining is the same and can be applied to either a tangible product or a service offering.

Summary

Six Sigma for Marketing and Six Sigma for Sales are relatively new approaches to enable and sustain growth. They are part of the bright future offered by adapting Six Sigma to the growth arena. The linkage of Six Sigma for Marketing and Six Sigma for Sales tasks and tools to strategic, tactical, and operational processes is where the Six Sigma discipline adds measurable value to marketing and sales team performance. Marketing and sales professionals can custom-design *what to do* and *when to do it* to fit these three critical marketing process arenas to their organization or culture. This book's concepts can complement your company's unique marketing approach and infrastructure. Why? Because the most important goal is to communicate a common approach to manage risk and make sound, data-driven decisions as you seek to expand the company. An organization can take license to customize the methodology to fit existing processes, enhancing communication and adoption. A customized application of this book's concepts will work as long as the following are upheld: phase objectives (or requirements), the sequence, tools-tasks-deliverables combinations, and phase-gate reviews. Integrating these methods and concepts into your critical processes with adequate rigor applied to meet deliverable requirements at phase-gate reviews will lead to more predictable outcomes.

Before exploring the details of each strategic, tactical, and operational method for marketing and sales, let's examine two foundational topics that transcend these three areas. The first fundamental subject involves the criticality of reporting and tracking performance and risk. Chapter 2, "Measuring Marketing Performance and Risk Accrual Using Scorecards," introduces a system of scorecards that build on Six Sigma principles to measure marketing's use of tools, completion of tasks, and the resulting deliverables across the strategic, tactical, and operational processes. Chapter 3, "Six Sigma-Enabled Project Management in Marketing Processes," addresses the importance of project management. We suggest adding some Six Sigma tools to the traditional project management body of knowledge to better manage a project and its associated risk.

2

MEASURING MARKETING PERFORMANCE AND RISK ACCRUAL USING SCORECARDS

A System of Scorecards to Measure Tools Usage, Task Completion, and Deliverables Across Marketing Processes

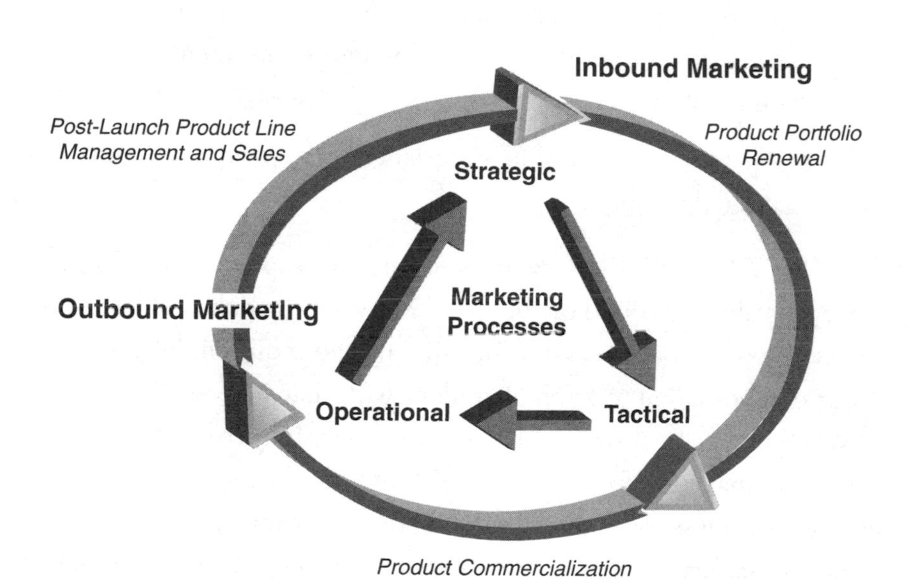

Scorecards in Marketing Processes

Similar to an adventurer's reliance on a compass, a businessperson's most powerful guiding tool to ensure that the business stays on course or adheres to its plan is a scorecard. A scorecard is the primary predictive tool for both in-process measures and performance results. Whether working in the strategic, tactical, or operational environment, marketing professionals need to take into account how to measure progress against their goals. Accountability within and across marketing teams is essential for proper risk management within a system of phases and gates (or post-launch milestones). As marketing work gets created and deployed, the issue of "pay for performance" also should be addressed. Thus, we have two distinct reasons to establish formal methods for measuring marketing performance:

- Manage risk and make key decisions at Gate Reviews and key project milestones to ensure adherence to the plan or to *explore* indications of critical changes requiring adjustments to stay on plan. To avoid going "off plan," we recommend making changes based on leading indicators to stay "on plan." Changes should be made to fulfill the plan rather than changing the plan too quickly. Our Six Sigma for Marketing tools and tasks are what change, *not* the deliverables and requirements, which are immutable.
- Pay for performance in accordance with specific requirements and deliverables.

Let's deal with the second topic first. We expect to be paid fairly for the value we add to the business. We want the deliverables we are working on to be successful. In fulfilling the requirements, we want our compensation to be in alignment with that success. But how do we pay fairly for projects that are unsuccessful for good reasons—reasons that data from our use of tool-task clusters helps substantiate? Business executives find it difficult to tell if they are fairly compensating (especially if projects are canceled) without a balanced system of scorecards. Scorecards should answer the following questions: "Is individual performance on track?", "Is the team's performance on track?", and "Are we all on target to meet our gate requirements by way of our deliverables?" These questions should be asked frequently to ensure that you make adjustments or cancel a project when the data suggests that you are trending off-track. Many managers see negative data and try to find a way to ignore its harsh implications. They press forward by using ambiguous requirements or often add "window dressing" when communicating the data to higher-level management. A system of scorecards can help address these concerns. Scorecards should track performance (against plan) and enable risk management (the accrual of risk over time). The system should include leading and lagging indicators that are critical to managing the business.

A tool is as good as the information it produces. The adage "garbage in, garbage out" could not be more applicable. Scorecards must track the right information to be useful to a data-driven marketing leader. Taking the time to determine the critical marketing risk accrual requirements defines the appropriate information to design into and track in a scorecard. The requirements may vary based on the tool, task or gate deliverable. Hence, a "system" is best. This chapter focuses on helping you address the key three requirement conditions to be successful using scorecards: using the *right tool* applied to the *right task at the right time* and delivering the right *summary data* for risk management and decision-making. Requirements are the questions asked *before* a phase of work is conducted. The goal is to design measurable work in light of the requirements *before* you start measuring.

Scorecards are all around you. They mean a variety of things to marketing professionals. The current literature represents a plethora of research on the best marketing scorecards. Topics include brand equity, customer equity, target markets, portfolio analysis and management, balanced scorecards, and return on investment (ROI) (and how best to apply it to marketing). Marketing scorecards typically include product market share, revenue growth from products, services, licenses, financing, and customer support. In addition to financial metrics, marketing scorecards include customer satisfaction. Some thought-leaders believe in the importance of a marketing event scorecard that tallies the investment (dollars and head count) against the number of participants, the change in awareness/perception/consideration levels, and the number of qualified prospects generated. Some marketing scorecards add the dimensions of awareness, image, perceptions, and consideration. These are all mainly lagging indicators and enable reactive behaviors. With Six Sigma thinking, this is necessary but insufficient.

The myriad of thought-leader recommendations or considerations for marketing scorecards are valid. Research undoubtedly will continue in this arena on how best to balance the scorecards. The literature is just beginning to introduce Six Sigma concepts to marketing scorecard thinking. We believe that regardless of the current thinking as to what should be measured, you should keep in mind some important Six Sigma concepts when building or revising your marketing performance scorecards. Our recommendations simply add more of a project focus to tracking marketing performance and risk management. We focus on finding and using leading indicators.

In general, we will discuss two types of tracking tools: checklists, as the simpler version, and the scorecard itself. More complex scorecards also can be referred to as dashboards, denoting that all the essential key indicators—the critical parameters—needed to "drive" a business are together in one spot.

Checklists

One traditional form of accountability is the checklist. Checklists assess two states of task completion—done or not done. Did you complete a certain task? Did you use this or that tool to enable the task? Checklists suffer from a lack of discrimination. They fail to provide information about quality or quantity. Checklists lack predictive specifics about how well a tool was used or a task's percentage of completion against its original target. Checklists are a good reminder tool rather than a predictive tool. Checklists prompt people to recall what needs to get done and track its completion. The checklist catalogs the expected requirements, tasks, or deliverables. Checklists monitor whether an item has been completed (a binary yes or no response) to help avoid duplication of effort. They fail to discriminate on the details of risk accrual.

Scorecards

Scorecards help scrutinize a business's health in terms of its performance and risk accrual. The scorecard drills down into the quality of processes or deliverables against respective targets and requirements. Scorecards can predict trends and alert businesspeople to potential changes that require a response to *maintain the planned course*. A system of scorecards presents those vital few indicators needed to understand how well a business is performing against its goals. A subset of scorecards probes deeper into an individual process or functional task/tool application area. Although this book's focus is marketing and its three processes (strategic, tactical, and operational), the concept of scorecards applies to any discipline to monitor the health of a task, project, process, or entire business.

This book uses a hierarchical flow of key criteria to describe accountability for completing the right things at the right time. Figure 2.1 shows the four criteria in sequence.

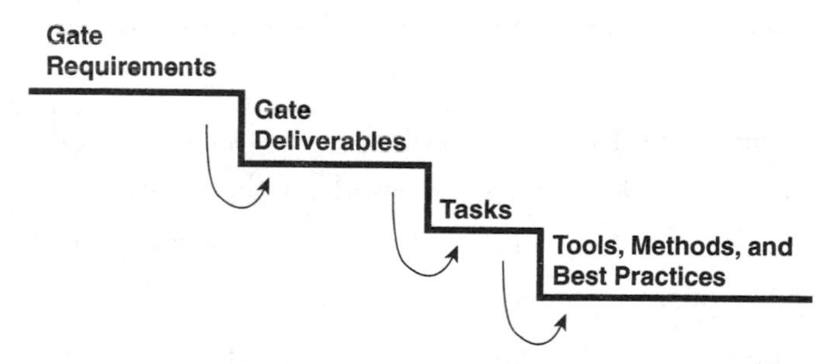

FIGURE 2.1 Hierarchical flow.

This four-level flow is well-suited to measurement by using an integrated system of scorecards. A system of scorecards can be designed and linked so that each subordinate level adds summary content up to the next level. This chapter discusses several types of scorecards for measuring performance and managing risk: tool scorecards, task scorecards, and two types of Gate Review scorecards.

Tool Scorecards

The most basic level of scorecard is filled out by marketing team members who use specific sets of tools, methods, and best practices to help complete their within-phase tasks. This is called a tool scorecard. It is easy to fill out and should not take more than 20 minutes or so to complete at the end of a tool application. Typically it is filled out in collaboration with the team leader for supervisory concurrence. After using a tool, the person or set of team members responsible for applying the tool to a task should account for the measurable items shown in Table 2.1.

TABLE 2.1 A Sample Tool Scorecard

Six Sigma Marketing Tool	Quality of Tool Usage	Data Integrity	Results Versus Requirements	Average Score	Data Summary, Including Type and Units	Task Requirement

The first column of the tool scorecard simply records the name of the tool used.

The Quality of Tool Usage column, as shown in Table 2.2, can be scored on a scale of 1 to 10 based on the following suggested criteria and levels. You can adjust these rankings as you see fit for your applications.

The integrity of the data produced by the tool usage can be scored using the suggestions shown in Table 2.3.

TABLE 2.2 Suggested Quality of Tool Usage Ranking

Rank	Right Tool	Fullness of Usage	Correct Usage
10	X	High	High
9	X	Medium	High
8	X	High	Medium
7	X	Low	High
6	X	Medium	Medium
5	X	Low	Medium
4	X	High	Low
3	X	Medium	Low
2	X	Low	Low
1	Wrong tool		

TABLE 2.3 Suggested Data Integrity Ranking

Rank	Right Type of Data	Proper Units	Measurable System Capability	% Data Gathered
10	Excellent	Direct	High	High %
9	Excellent	Direct	High	Medium %
8	Excellent	Direct	Medium	High %
7	Good	Close	High	High %
6	Good	Close	Medium	Medium %
5	Good	Close	Medium	Low %
4	Weak	Indirect	Medium	High %
3	Weak	Indirect	Low	Medium %
2	Weak	Indirect	Low	Low %
1	Wrong	Wrong	None	N/A

You can adjust the nature of the scoring criteria as you see fit for your applications. The key is to clearly delineate between various levels of measurable fulfillment of the criteria. This scoring stratification transforms business requirements into actionable performance criteria, thereby clarifying expectations.

The ability of the *tool results to fulfill the task requirements* is scored with the help of the following criteria:

- **10:** The results deliver *all* the data necessary to completely support the fulfillment or lack of fulfillment of the task requirements.
- **9 and 8:** The results deliver a *major* portion of the data necessary to support the fulfillment or lack of fulfillment of the task requirements.
- **7 through 4:** The results deliver a *moderate* amount of the data necessary to support the fulfillment or lack of fulfillment of the task requirements.
- **3 through 1:** The results deliver a *very limited* amount of the data necessary to support the fulfillment or lack of fulfillment of the task requirements.

This rating scale accounts for how well your data fulfills the original requirements. It is acceptable to find that a full set of data was in hand to determine that you cannot meet the requirement a task was designed to fulfill. Marketing professionals should be rewarded for doing good work. Even when the results are bad news, they communicate the truth. The intent is to avoid false positives and false negatives when making decisions about a project's viability. This metric helps quantify the underdevelopment of data and facts that can lead to poor decisions.

Task Scorecards

Task scorecards can evaluate performance relative to its requirements at both the aggregate level of tool completion and the summary level for each major task. Table 2.4 shows a sample task scorecard.

The average tool score is simply the averaging of tool scores that are aligned with each major task. A very insightful metric for each major task within a phase is the percent complete or percent task

TABLE 2.4 A Sample Task Scorecard

Phase Task	Average Tool Score	% Task Fulfillment	Task Results Versus Deliverable Requirements	Red	Yellow	Green	Deliverable Requirements

fulfillment. We believe this is where the real mechanics of cycle time are governed. If a marketing team is overloaded with projects or is not given enough time to use tools, it is almost certain that they will not be able to complete their critical tasks. Undercompleted tasks are usually a leading indicator that you are likely to slip a schedule. Too few tools are being used, and the ones that are being used are not being fully applied. So there is a double effect—poor tool use, leading to incomplete data sets, and tasks that simply are unfinished. The average tool score also tends to be low. This means that you make risky decisions on the wrong basis. It is fine to face high-risk situations in your projects, but not because you are too busy to do things right. Task incompletion is a major contributor to why you make mistakes and fail to grow on a sustainable basis. The following list is a suggested ranking scale to illustrate how you can assign a value from 1 to 10 to quantify the level of risk inherent in the percent of uncompleted tasks:

- **10:** The task is complete in all required dimensions. A well-balanced set of tools has been fully applied to 100% completion.
- **9 and 8:** The task is approximately 80 to 90% complete. Some minor elements of the task are not fully done. A well-balanced set of tools has been used, but some minor steps have been omitted.
- **7 through 4:** The task is incomplete, somewhere in the range of 40 to 70%. Moderate to major elements of the task are not

done. Tool selection and usage have been moderate to minimal. Selected tools are not being fully used. Significant steps are being skipped.

- **3 through 1:** The task is incomplete, somewhere in the range of 10 to 30%. A few tools have been selected and used. Their steps have been heavily truncated. Major steps are missing.

The column comparing the task results versus gate deliverable requirements identifies how well the work satisfies project requirements. If you complete 100% of the critical tasks and use a balanced set of enabling tools to underwrite the integrity of your deliverables, you are doing your best to control risk. You can produce outstanding deliverables, full of integrity and clarity, that fail to meet the requirements for the project and its business case. You have great data that tells you that you cannot meet your goals. This is how a gatekeeping team can kill a project with confidence. Not many companies kill projects very often, and even fewer do it with tool-task-deliverable confidence.

You must consider two views when managing risk and making gate decisions:

- Fulfilled gate requirements, indicated by a positive "green light" to continue investing in the project.
- Unfulfilled gate requirements, represented by a cautioning "yellow light" or a negative "red light" that signals a redirecting of the project or an outright discontinuance of the project.

A color-coded scheme of classifying risk can be defined as follows:

- **Green:** 100% of the major deliverables are properly documented and satisfy the gate requirements. A few minor deliverables may be lagging in performance, but they present no substantive risk to the project's success on three accounts: time, cost, and quality.

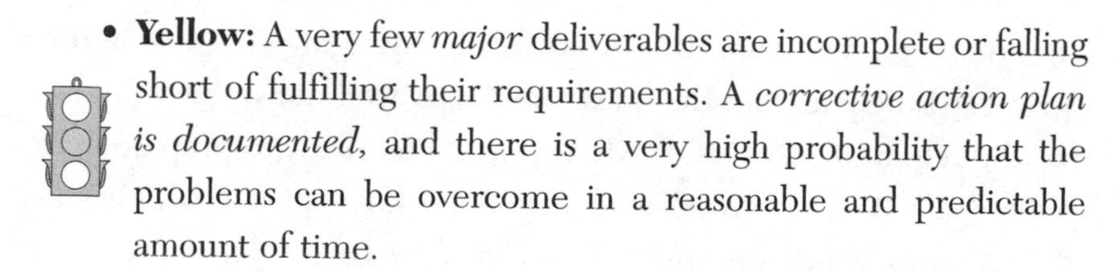

- **Yellow:** A very few *major* deliverables are incomplete or falling short of fulfilling their requirements. A *corrective action plan is documented,* and there is a very high probability that the problems can be overcome in a reasonable and predictable amount of time.

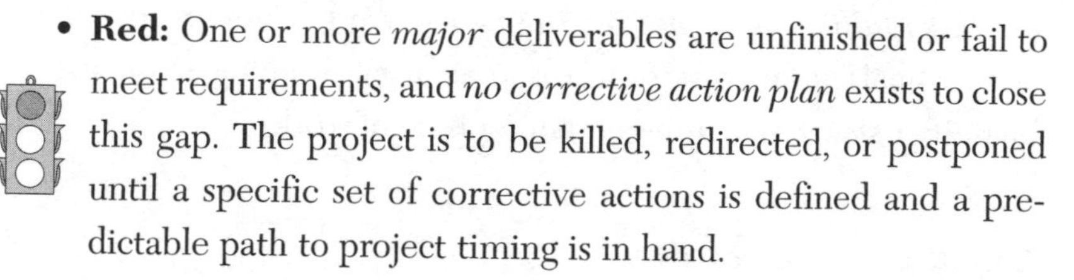

- **Red:** One or more *major* deliverables are unfinished or fail to meet requirements, and *no corrective action plan* exists to close this gap. The project is to be killed, redirected, or postponed until a specific set of corrective actions is defined and a predictable path to project timing is in hand.

The following is a suggested set of ranking values to quantify the risk associated with varying levels of mismatch between the gate requirement and what is delivered from a major task:

- **10:** The results deliver *all* the data necessary to completely support the fulfillment or lack of fulfillment of the gate requirements.
- **9 and 8:** The results deliver a *major* portion of the data necessary to support the fulfillment or lack of fulfillment of the gate requirements.
- **7 through 4:** The results deliver a *moderate* amount of the data necessary to support the fulfillment or lack of fulfillment of the gate requirements.
- **3 through 1:** The results deliver a *very limited* amount of the data necessary to support the fulfillment or lack of fulfillment of the gate requirements.

To recap, we have demonstrated how a tool scorecard documents the quality of tool use, the data's integrity, and the fulfillment of a task requirement. We have gone on to the next level of scoring risk by defining a task scorecard. Here project leaders can quantify how well one or more enabling tools have contributed to completing a major task, what percentage of the task has been completed, and how well the deliverables from the task fulfill the gate requirements. Scorecards help answer the questions "How well did we do in meeting the requirements?" and "Are we done and ready to prepare for a phase Gate Review?" A positive answer to these questions indicates the following conditions: *the tools fulfilled the task requirements* and *the tasks fulfilled the gate requirements*. We are now ready to look at the final summary scorecards that a gatekeeping team uses to quantify risk accrual at the end of a phase of work.

Gate Review Scorecards

Gate review scorecards are used to assess accrued risk and make decisions at each gate review or major milestone. The two kinds of Gate Reviews are functional level Gate Reviews and executive level Gate Reviews.

- **10** = Results deliver *all* data necessary to completely support the fulfillment or lack of fulfillment of the Task requirements.
- **9 through 8** = Results deliver a *major* portion of the data necessary to support the fulfillment or lack of fulfillment of the Task requirements.
- **7 through 4** = Results deliver a *moderate* amount of the data necessary to support the fulfillment or lack of fulfillment of the Task requirements.
- **3 through 1** = Results deliver a *very limited* amount of the data necessary to support the fulfillment or lack of fulfillment of the Task requirements.

Functional reviews are detailed and tactical in nature and prepare the team for an executive review. An executive review is strategic in nature and looks at macro-level risk. Risk management at the executive level delves into how a particular project contributes to the overall portfolio of commercialization projects or how it manages risk in the post-launch environment. Thus, functional gatekeepers worry about micro details within their particular project. Executive gatekeepers worry about accrued risk across all projects that represent the business's future growth potential *as a portfolio*. The former looks at alignment of risk within the specific project's tactics, and the latter looks at alignment of project risk across the business strategy. Functional reviews can be done for technical teams and marketing teams as independent events. Executive reviews are summary presentations that should integrate both technical and marketing perspectives, as well as any other applicable macro-gate deliverables.

Table 2.5 shows a generic template for a functional Gate Review.

TABLE 2.5 A Sample Functional Gate Review Scorecard

Gate Deliverables	Grand Average Tool Score	Summary of Task Completion	Summary of Task Results Versus Deliverable Requirements	Red	Yellow	Green	Accountable	Target Completion Date	Corrective Action and Risk Assessment

The integrated system of tool, task, and gate deliverable scorecards provides a *control plan* for quantifying accrued risk in a traceable format that goes well beyond simple checklists. Control plans are a key element in Six Sigma. The task scorecard feeds summary data to this functional form of a gate deliverable scorecard.

The next Gate Review format is used for *executive Gate Reviews*, where deliverables are summarized for rapid consumption by the executive gatekeeping team (see Figure 2.2).

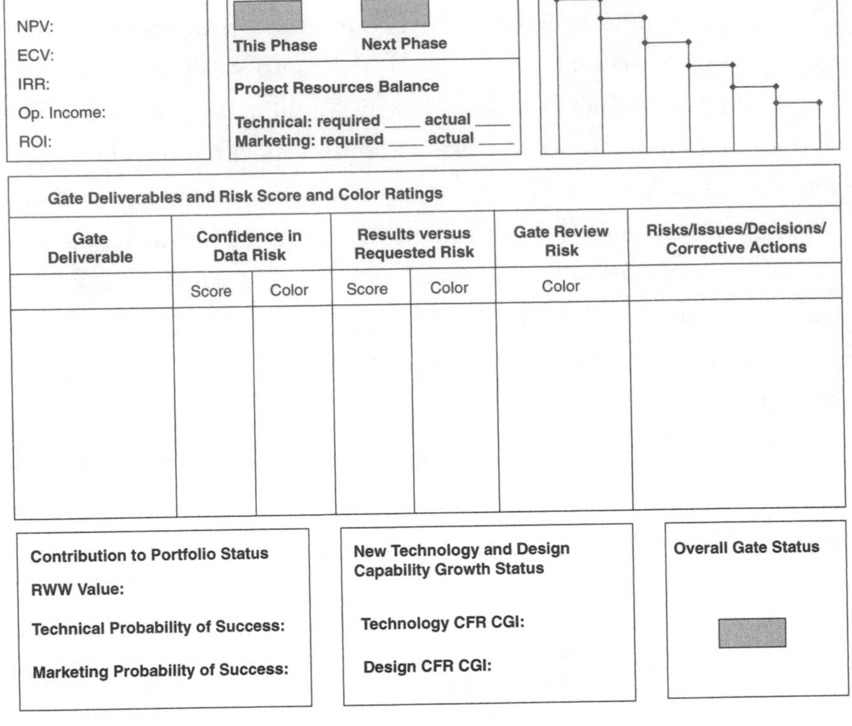

FIGURE 2.2 A sample executive Gate Review scorecard.

This template is a common format employed by numerous companies that have a commitment to strategic portfolio management. The executive gatekeeping team looks at a number of these scorecards to balance risk across its portfolio of projects while driving growth to the top line in accordance with the business strategy.

Conducting Gate Reviews

Conducting a Gate Review requires forethought and planning on the part of both the project team members and the reviewers. The project members' planning entails creating the meeting agenda; distributing the pre-read; and gathering, analyzing, and summarizing the project status in an appropriate presentation review format. The project members also need to pay close attention to *who* can participate in the review. Often who attends is as important as the content.

At the beginning of the project, the sponsor(s) and the project manager ensure that the right players and the right roles are invited to participate in project reviews. If a project has multiple sponsors or multiple key management stakeholders, having them together in one forum to discuss and debate issues, implications, and impacts often yields different results than one-on-one discussions. The interchange among the key players in a single forum helps validate the legitimacy and feasibility of risk management concepts that need cross-functional scrutiny before being implemented. If more than one key person is unable to attend, consider rescheduling the meeting.

Another easily overlooked responsibility of the project team is the *timing* of the reviews. The timing of the Gate Review should correspond with the project content. Given management's often-busy schedule, there is a tendency to put review dates on the calendar months in advance, sometimes through the project's anticipated duration. However, project reviews should be conducted upon completion of key milestones and/or deliverables. During the project startup phase, the preplanning work involves defining and clarifying customer requirements (both internal and external). Those requirements then get translated and parsed into supplier (or technical) specifications. Next, each specification becomes a deliverable and gets assigned to the person accountable for producing the deliverable. The necessary activities to complete that deliverable get defined. The appropriate support tools are identified. The person accountable for producing the deliverable commits to a target completion date. Finally, all these steps are summarized in a project plan according the agreed-on methodology. Simulating project schedule options helps create awareness of the *potential range* of timing outcomes. All these steps leading up to and including the project plan should inform the scheduling of the Gate Review meetings. Rather than having the calendar determine the timing of the review, the completion of gate deliverables and the availability of the appropriate participants should dictate the review date.

The Gate Review participants should include only the critical project team members and the reviewers. The reviewers are the sponsors and key stakeholders. The sponsor pre-meeting planning responsibilities start at the beginning of the project with the selection and invitation of the Gate Review participants. Ensuring the appropriate representation and mix of roles is vital to guiding the project to successful completion. The main objective is proper gathering, prioritization, and communication of project requirements. To achieve that objective, the sponsor and key stakeholders determine the appropriate balance of customer requirements with business requirements. In addition, functional or organizational ownership across the value chain must be established. Allowing the value-chain stakeholders to help shape the project requirements and/or translation into specifications (deliverables) early on in the project life helps build a sense of ownership and commitment. Value-chain commitment is critical to implementation success. Hence, ensuring appropriate involvement of critical value-chain roles in Gate Reviews or participation as project members is a fundamentally important task for the sponsor.

The objective of a Gate Review is to assess and hopefully approve the completed deliverables against the requirements. The sponsor needs to clearly define requirements such that they are unambiguous. An integrated set of internal and external requirements is the criteria against which a project's work is evaluated. The expectation setting for each Gate Review is also the sponsor's responsibility. Ideally, Gate Reviews *avoid* being a mystery or take on an "investigative" tone wherein "whodunits" are sought. Gate Reviews are an opportunity to guide and direct the project, rather than letting the team figure it out on their own. If a course correction is needed, it can be done within predetermined criteria under the project and risk management activities. Upon completion of one review, the sponsor should reinforce and clarify what the success criteria are for the next review. The agreed-on methodology and project plan should inform the sponsor when defining and articulating the next-step expectations to the project team.

In addition to confirming next-phase project requirements within each Gate Review, the reviewers also need to inspect the completed and in-process deliverables. The reviewers must judge whether the completed deliverables meet their respective requirements. One way is to ask which tools were used to achieve the task. Understanding if the appropriate tool was used for the right purpose to answer the right question provides insight into any potential project team assumptions or leaps of faith made to complete the task. If the incorrect tool was used, more than likely the wrong question was asked, or the answer doesn't match the question. A summary of the tool scorecards should be available for inspection if additional insight is required at a functional review. Specific points of interest are the quality of tool usage and data integrity.

The sponsor and project manager share a Gate Review responsibility—*communication*. Communication spans the project and meeting life cycle. Successful communication results from thoughtful and thorough preplanning and skilled execution. The preplanning spans the "who, what, when, where, why, and how" test—*who* needs to receive *what* information, and *when*, *where* (the address), and *how* they prefer to receive the information. The *why* shapes the communication message—Why does this person care about the information? (And what's in it for me? [WIIFM].) Succinct, actionable, informative communications about the project help keep the target audience engaged and interested. How the message is communicated—in what format (written, video, voice) and in what tone (direct, entertaining, informative)—has to match the audience's preference and the message's intent. You also should consider *how the target audience can ask clarifying questions* to ensure that they have received the message as intended. Communication is one of the most important responsibilities of the sponsor, reviewers, and project members. Whether the communication takes place in a Gate Review or before or after that, the importance of communicating well is tantamount.

Summary

You have built an integrated system of scorecards that can be modified to suit any organization's phase-gate process (see Figure 2.3).

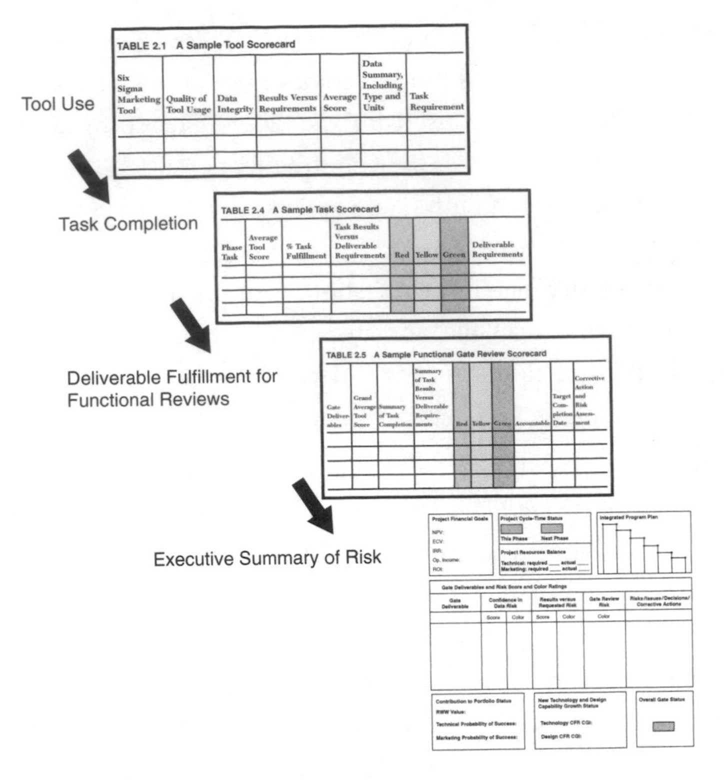

FIGURE 2.3 Scorecard summary.

Each hierarchical level should feature a scorecard designed to mirror the respective business model from one of the following areas:

- Your tool-task groups (clusters)
- Your deliverable-gate requirement groups as they are aligned with the phases of your portfolio renewal process
- The product commercialization process
- Your post-launch line-management process

We have provided examples of a scorecard system for each of the process arenas (strategic, tactical, and operational).

The scorecard system can be used to manage the portfolio renewal process. The first three scorecards are sufficient to manage risk within a project. As each product commercialization project is activated from the portfolio renewal team, an executive summary scorecard is initiated and updated during commercialization.

As the old saying goes, "That which gets measured—gets done." We discourage monitoring performance with checklists; they serve as reminders of what to use or complete, *not* how well that expectation is being met. We highly recommend the harder, but more responsi ble, approach of the scorecard system. A custom-designed scorecard set, spanning the four-level hierarchy of tools-tasks-deliverables-requirements, serves as both a quantitative and a qualitative barometer. Scorecards identify not only *what* to measure (the critical performance elements) and *how well* these critical parameters satisfy a set of criteria (or requirements), but also *when* to measure them so that you have a much higher probability of preventing downstream problems as you seek to sustain growth across your enterprise.

Now let's explore the importance of project management. We suggest adding some Six Sigma tools to the traditional project management body of knowledge to better manage a project and its associated risk. Chapter 3, "Six Sigma-Enabled Project Management in Marketing Processes," features specific recommendations on how certain Six Sigma tools enhance project management.

3

Six Sigma-Enabled Project Management in Marketing Processes

Designing Cycle-Time for Strategic, Tactical, and Operational Marketing Projects

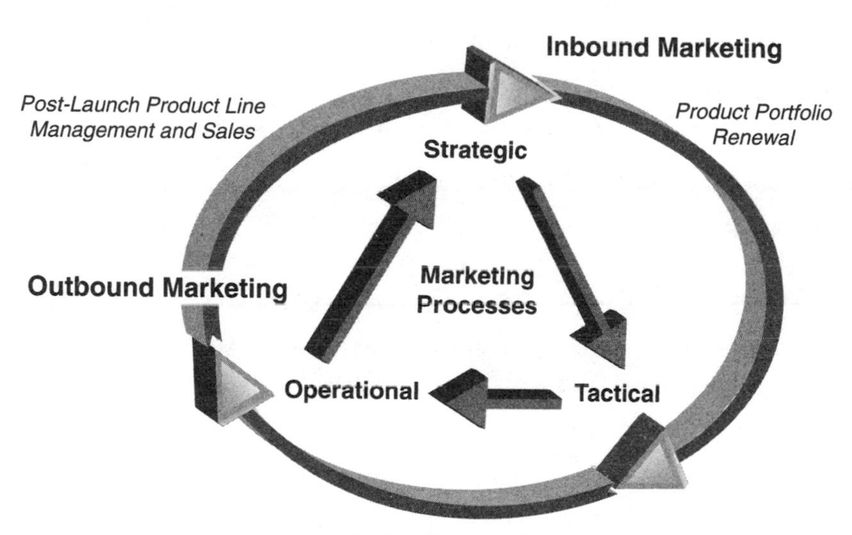

Six Sigma Contributions to Project Management in Marketing Processes

A great deal of standard material is available to help marketing professionals generate a project plan. It is not our intent in this chapter to review the basics of the project management body of knowledge. Our goal is to demonstrate how a few value-adding elements from traditional Six Sigma tools help in the design and analysis of marketing cycle time. We want you to have high confidence that what you choose to do and how long you forecast that it will take are in alignment with management expectations.

We have found, over the last decade of studying project cycle time, that executives and work teams are really out of touch with one another. This is true of both marketing and technical teams across the strategic, tactical, and operational processes. Executives have a cycle-time expectation that is almost always dramatically shorter than what their teams say they need to do the job right. Completing a set of tasks correctly and fully is what teams try to do. Management forces these teams to do their work faster than that. Why does this happen? Usually the key reasons stem from the business's doing too many projects and from being unaware of when and why each project should finish. The thinking is "the faster we get done, the sooner the cash will flow in, and we hope we will make our numbers." Busy people present a comforting image that they are helping increase income.

This rushed, frenetic, short-term-focused behavior causes three consequences: unfinished critical activities (or tasks); incomplete key deliverables; and ad hoc usage of tools, methods, and best practices. Any combination of these three can result, but they merely serve as a warning sign of unsustainable growth. Rushing works, but not on a sustainable basis. Rushing produces sporadic growth. If you want consistent growth, allow your teams the time they need for proper planning and completion of requirements (using the appropriate tools-tasks-deliverables combination).

Unfinished activities or tasks occur when a flow of work stops prematurely—because of lack of time, lack of funds, or superseding priorities. For example, completing a task may require seven essential steps; however, shortcuts taken in the process may have truncated some of the steps. If most senior executives assessed task completion below a superficial look, they would be surprised by the number of shortcuts that occur in their company. Our collective experience tells us that this "abridged" approach negatively affects your ability to meet your growth goals. Too often executives are unaware of its occurrence or its magnitude of effect on the business. Yes, work gets done, but it gets done incorrectly and incompletely. Work gets done just to the point of "that's good enough for now—we'll clean it up later when we have time" (which seldom happens, because we have too many other projects to do so that we can make our growth numbers). Growth is not sustainable under these self-defeating conditions.

Incomplete deliverables wield a similar, but more obvious, outcome than unfinished tasks. At the juncture of a Gate Review, the data needed to assess risk and make key decisions either is absent or is less helpful to the executives who confirm that the phase-gate requirements have been met. Summary data gets trumped up to appear to have enough credibility to enable the gatekeepers. People can get clever at dressing up incomplete data to look good. The typical gatekeeper, when presented with this situation, immediately goes into "best judgment" mode. This tactic saves face but cannot replace balanced sets of data that tell the truth. Taking incomplete sets of summary data and filling in gaps with judgment is routine in modern corporations. When this becomes a standard practice, sustainable growth is not. You have a mismatch between what data is required to make decisions and what data your teams are allowed (funded and expected) to produce.

Clients tell us that another fallout of an environment that breeds incomplete data is that the talented people shy away from such projects. The experts opt out of projects that expect them to sell half-baked data at Gate Reviews because they find them unrewarding. Hence, less-experienced people staff these projects, thereby causing learning curve issues when producing complex deliverables. We have observed that under pressure, management tends to load teams to the hilt, even with B-grade projects. Chapter 4, "Six Sigma in the Strategic Marketing Process," shows how to cut these projects back to an acceptable level.

Ad hoc use of tools, methods, and best practices yields inconsistent results. A compounding effect starts with misaligning tasks with their enabling tools. The tools become viewed as unimportant or things that slow down progress. Unimportant tools become ignored tools. Ignored tools obviously cannot fulfill their purpose of generating consistent and proper results. Lack of a systematic linkage of enabling tools applied to completing a task (or a series of related tasks) produces unpredictability. Critical tasks get done in any number of ways. In fact, an undisciplined application of enabling tools can elicit ill-defined tasks and incorrect data summaries. Often personal preference determines which tools get used (the familiar or "easy" tool), as opposed to a standard of excellence that is universally recognized by management. Hence, the individual deciding how to complete a task ignores best practices (standard work) that shape the adoption of supporting tools and methods. Consequently, inconsistency defines the standard for unchecked application of tools, methods, and best practices.

There is much you can do to fix this mismatch. A disciplined approach for both management and the work teams can stabilize inconsistencies and drive predictable results. The project management discipline coupled with Lean principles and Six Sigma can better design and statistically model cycle time. This combination can improve understanding and documentation of critical path failure modes and then better resolve which tasks and enabling tool sets are critical to producing the necessary deliverables for proper decision-making at a Gate Review or project milestone meeting. When such work is designed, funded, and expected, risk is much easier to manage.

Designing Cycle Time: Critical Paths of Key Marketing Task-Tool Sets by Phase

Designing cycle time is a nine-step approach. It differs from simply creating a Work Breakdown Structure (WBS) and setting up a Gantt chart (to show how long it takes to complete a series of tasks). The unique, designed cycle-time characteristics include the following:

- **Gate requirements** are used to define very specific gate deliverables.
- **Gate deliverables** are summary results that come directly from within-phase tasks that are done fully and completely. (What is worth doing is worth doing right.)
- **Within-phase tasks** are underwritten by a linked set of enabling tools, methods, and best practices that are proven to be effective in developing the data required for the gate deliverables.
- **Tool-task-deliverable sets** are linked for maximum value in fulfilling gate requirements (think "Lean" in this context). (The term "Lean" references only those *vital few* tool-task-deliverable sets needed to meet the gate requirements.)
- **Balanced and trained resources** can be properly aligned with the detailed tool-task-deliverable sets, identifying who and what skill set you need to develop the gate deliverables—the *core competencies*!

Designing a critical path of cycle time with these five characteristics greatly increases the likelihood of meeting gate requirements. This represents a key step on the path to sustaining growth. With this structured approach, projects will be identified as deserving either investment to sustain the project to meet growth goals or termination. Real-winning-worthy projects from your activated portfolio of new or existing offerings (products and services) will stand out. Decisions to fund these projects can be made with certainty. Conversely, projects with too many critical gate deliverables, or those that fall short of meeting the gate requirements, can be killed with confidence. You know that you are doing the right thing for the sake of the business. "Designed" cycle time provides the assurance that a given project's potential is leveraged and optimized.

The following list describes the nine key steps to designing cycle time correctly. This list requires a lot of discipline and tenacity to complete. However, our clients tell us the benefits far outweigh the investment. Those who reliably execute these nine steps attain their desired growth goals repeatedly over time; conversely, those who give up are unable to sustain growth. Most of these steps come from the project management discipline. However, four of the steps are derived from the Six Sigma discipline: Steps 3, 4, 8, and 9:

1. Define gate requirements clearly and completely.
2. Define gate deliverables clearly and completely.
3. Develop detailed process maps, WBSs, and workflow charts.
4. Link major tasks to balanced sets of tools, methods, and best practices.
5. Define roles and responsibilities clearly and completely.
6. Develop a Pert Chart for each phase of the project.
7. Calculate the project's critical path.
8. Conduct a Monte Carlo Simulation to forecast the project's completion.
9. Create Failure Modes & Effects Analysis (FMEA) for the tasks on the critical path.

Let's go into more depth for each step:

1. **Define gate requirements clearly and completely.** Determine and document exactly what goals and targets you need to get through the gate. Identify what you need to manage risk and make key decisions.

2. **Define gate deliverables clearly and completely.** Document the summary data sets that represent the progress against the gate requirements. The truth is in this data (*not* opinions or guesses).

3. **Develop detailed process maps, WBSs, and workflow charts.** These three (Six Sigma-derived) tools show how a marketing process has controllable and uncontrollable inputs and outputs. They depict the flow of critical marketing tasks within each phase—a system of integrated work.

4. **Link major tasks to balanced sets of tools, methods, and best practices.** These represent a value-adding approach similar to DMAIC Six Sigma projects. Enabling tool sets (composed of methodologies and built with appropriate best practices for your industry and offering) defines how tasks really produce data-based deliverables. They ensure that the right data is collected and analyzed to properly fulfill a given task.

5. **Define roles and responsibilities clearly and completely**. Using the RACI Matrix defines individual accountability for task-tool completion. RACI defines who is **R**esponsible for doing the work, who is **A**ccountable for the people doing the work, who should be **C**onsulted to help get the work done, and who must be **I**nformed of task progress and results for parallel or dependently related tasks. It is important to note that only *one* person can be accountable for a task or activity; however, responsible, consulted, and informed can be represented by more than one person. In fact, a person can have multiple roles. For example, the person held accountable for a task or deliverable may also be held responsible.

   A Gantt Chart of tasks aligned in a what-who matrix expands its purpose to include informing the team as to what team members are accountable for and when. These tools prevent miscommunication.

6. **Develop a Pert Chart for each phase of the project.** A Pert Chart illustrates a network of serial and parallel flows of tasks as they occur sequentially. This visual flow of work shows task relationships and dependencies within a phase.

7. **Calculate the project's critical path.** The critical path identifies the longest critical tasks timeline (between starting and ending) in a phase. These tasks define your cycle time for a phase.

8. **Conduct a Monte Carlo Simulation to forecast the project's completion.** This Six Sigma tool identifies value-adding variables (tasks) by modeling. Statistical analysis and forecasting determine which selected variables (tasks) affect outputs (time) and quantify the effect's magnitude. (Often this input-output relationship is defined as a mathematical equation: $\Delta Y = f(\Delta X)$.) A Monte Carlo Simulation produces a frequency distribution that predicts the likelihood of finishing the critical path of tasks.

   Several good, user-friendly Monte Carlo Simulation packages are available. They include Crystal Ball by Decisioneering®, Inc.; Decision Pro by Vanguard Software Corporation; @Risk by Palisade Corporation; and Decision Making with Insight by AnalyCorp. Any software package can do the job. You can select the most appropriate tool based on the potential number of users, user interface, scope of analysis, and price. The simulation results then can inform your project schedule of tasks and milestones (that is, Microsoft Project). The @Risk software, for example, enables you to directly port its results to a Microsoft Project file.

9. **Create a FMEA for the tasks on the critical path.** FMEA is a very common tool used in most Six Sigma projects. This tool gives a detailed assessment of what can go wrong for each task on the critical path and the likelihood that it will occur.

FMEA also includes a prevention plan step to help ensure that tasks stay on track as much as possible and a response plan for when the risk is realized (a risk-mitigation plan).

Quality execution of these nine steps affords project teams more credibility to deliver their commitments at the quality promised, on time, and within budget. However, more than likely they will take longer to complete the project. To improve the adoption rate of this approach, our successful clients start by implementing each of the nine cycle-time design steps one by one and see how consistently growth goals are met year after year. This gives the project teams time to get comfortable with using the tools and introduces management to their benefits. By about the third project, they have replaced the ineffective deterministic (top-down) project deadline approach with the approach just described, and they enjoy the confidence of achieving goals right the first time. They become predictable.

Modeling Marketing Task Cycle Time Using Monte Carlo Simulations

As mentioned, four of the nine items are Six Sigma-based and go beyond traditional baseline project management methods. These four add incremental value in producing high-integrity project cycle time planning and results. Conducting a series of Monte Carlo Simulations on the designed critical path of tasks is a critical Six Sigma tool. The tool's importance justifies its own section consisting of eight key elements.

A Monte Carlo Simulates the amount of time to complete the critical path, depending on the amount of time each task takes. The method calls for changing (or varying) the time allocated for each task (ΔX) on the critical path, using carefully selected limiting conditions. Then the model adds up the entire time for the phase to be completed (ΔY) and provides a range of predictions with varying confidence intervals. The value-adding elements of a Monte Carlo Simulation on designing cycle time are as follows:

1. A *simple math model* that adds up each task duration for the critical path is defined as Y = f(X), where Y is total cycle time and each X is a task duration on the critical path (see Figure 3.1).

$$\Delta Y = \Delta X_{task\ 1} + \Delta X_{task\ 2} \ldots + \Delta X_{task\ n}$$

FIGURE 3.1 The Y = f(X) model.

2. Each task is represented as a *triangular distribution* of possible durations, typically in units of days or hours (ΔX) (see Figure 3.2). Possible task durations are statistically represented using a triangular distribution that illustrates a linearly diminishing likelihood of a task duration occurring at the ends of the triangle. The triangle's "peak" represents the most likely time a task will take to complete.

Example of Crystal Ball MC Software from Decisioneering

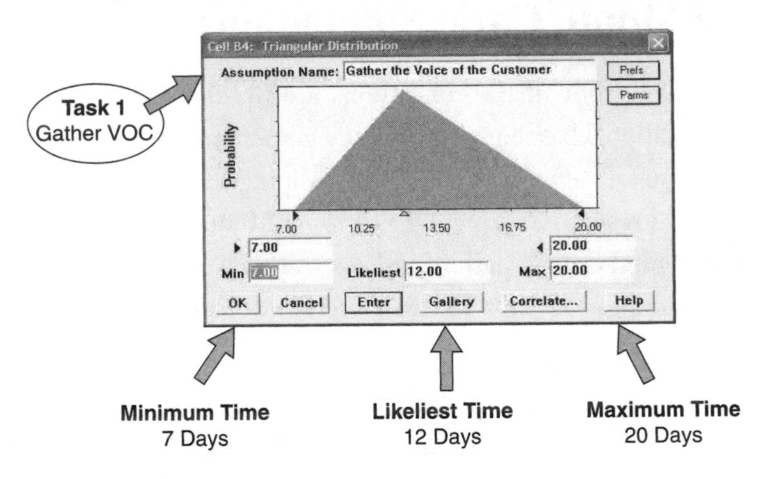

FIGURE 3.2 Crystal ball triangular distribution.

3. Each triangular distribution is "loaded" with your *best estimates of three types of durations* that can be projected for each task. These estimates must come from a team of marketing veterans

who have experience in how long tasks take to complete using the appropriate tools. "Garbage in, garbage out" can take hold here if you aren't careful; we prefer to work with $\Delta Y = f(\Delta X)$. Know your tools and how long they take to enable the completion of tasks. Triangular distributions provide samples of data to calculate task completion time three different ways:

The *shortest likely time* to complete a task (when things go really well—yeah, right!).

The *most likely time* (the mean or median) to complete a task.

The *longest likely time* to complete a task (when things go really wrong—Murphy's Law awaits!).

4. The Monte Carlo Simulation randomly samples or selects a task-completion duration from within the triangular distribution, representing each task's possible range of durations. The software running the simulation uses *a random-number generator* to "grab" an unbiased duration from within each triangular distribution.

5. The sum of each run through the critical path of randomly selected task durations is calculated and entered into a histogram (frequency distribution), as shown in Figure 3.3.

6. The final frequency distribution estimates a range and frequency of distributed cycle-time end points for the phase. The curve's final shape (usually a reasonably normal distribution) and its statistical parameters are graphically presented for analysis. (The team should use this final curve to have an enlightened discussion about the reality of their estimated completion time.)

7. The *percent confidence* in finishing the project on or before a specific date can be viewed on the frequency distribution chart. If the simulated answer seems unacceptable, the critical path can be redesigned until a compromise is reached on what tasks to perform, when to do them, and how long each task must take

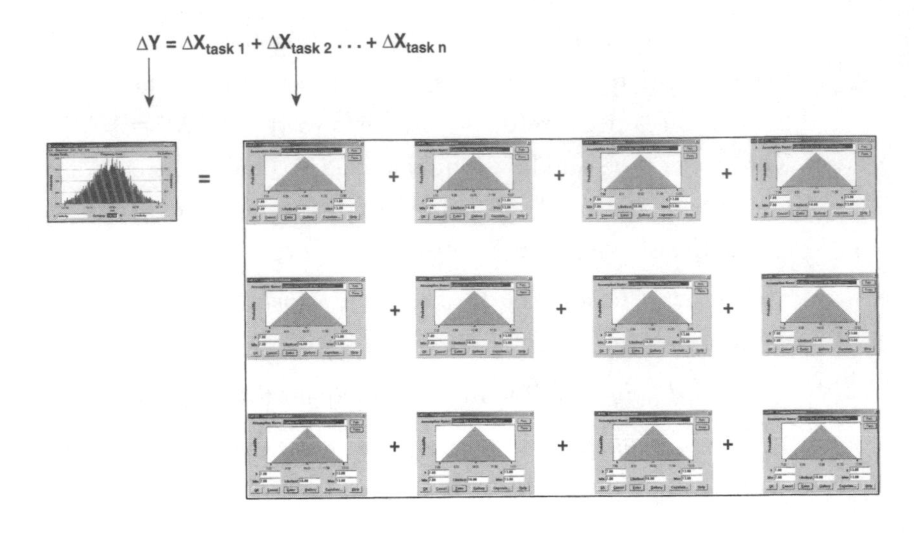

Example of Crystal Ball MC Software from Decisioneering

FIGURE 3.3 The sum of each crystal ball run.

(on a task-by-task basis). Reduce the most likely outcome (the mean or median of the cycle-time frequency distribution [Y]) by reducing the mean or median duration values of each task (Xs). The range of the forecast distribution (or width) can be narrowed by changing the values of the shortest and longest likely times for each task in the critical path (see Figure 3.4).

Referencing historical data can help the team avoid tweaking the inputs simply to deduce the desired output.

8. Use the FMEA to help rationalize changes made to the tool-task sequences that deliver the final cycle-time risk you can bear. Decide what tools, methods, and best practices are must-haves versus ones that are optional (lean out the cycle time). Decide what tasks are absolutely critical to be done fully and completely versus those that can be done with lesser levels of completeness (more leaning out the cycle time). Finally, a "leaned-out" sequence of the critical tasks and their enabling tools will define the critical path. This is Lean Six Sigma applied to project management.

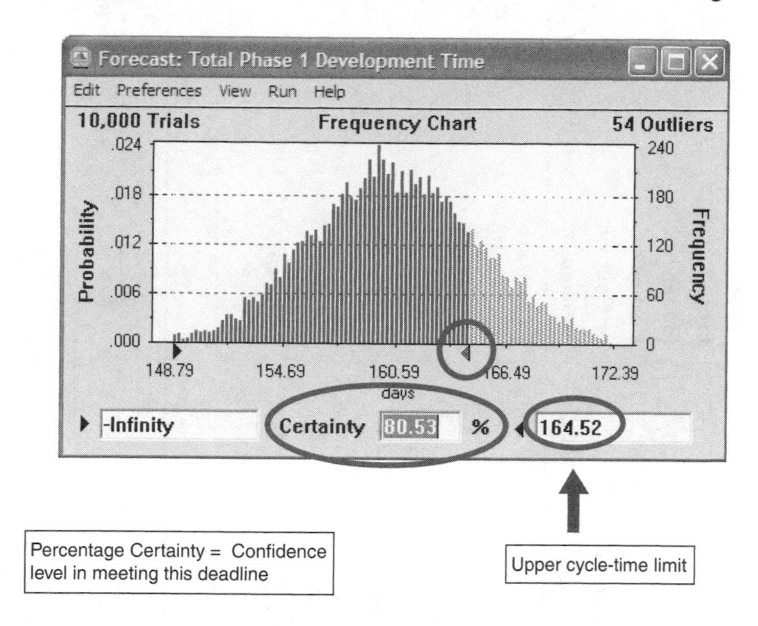

FIGURE 3.4 Forecast distribution of critical path shortest and longest times.

Documenting Failure Modes in the Critical Paths of Marketing Tasks by Phase

The last unique, value-adding method for traditional project management is applying FMEA to the cycle-time design. This is one item you almost never see used in project management, yet it is the easiest to apply, and you get a really good payback for the time invested in doing it.

FMEA can be applied to any process; it works great to help identify risks and respective consequences. Using the FMEA tool helps teams design a proactive approach to avoiding cycle-time problems—a key enhancement. Some people may use FMEAs to support developing reaction plans to address problems if and when they arise, but that is not the most effective way to use the tool. A risk analysis and mitigation tool from the project management discipline can be just as effective in contingency planning. Table 3.1 shows a classic FMEA template.

TABLE 3.1 Project FMEA

Task/Tool	Tool Function	Potential Failure Mode	Potential Failure Effects	SEV	Potential Causes	OCC	Current Project Management Evaluation or Control Mechanism	DET	RPN
What is the Task/Tool application under evaluation?	*What is the purpose of the tool?*	*In what ways does this Task/Tool application corrupt cycle-time goals?*	*What is the impact on the project cycle time?*	*How severe is the effect on the schedule?*	*What causes the loss of Task/Tool function?*	*How often does cause or FM occur?*	*What are the tests, methods, or techniques to discover the cause before the next phase begins?*	*How well can you estimate cause?*	
									0
									0
									0
									0
									0
									0

As shown, a project FMEA has seven basic elements and three quantifiable attributes to calculate the magnitude of the potential risks (as denoted by the column headings). First, let's examine the seven basic FMEA elements:

- **Task/Tool:** Identify the task (function) or tool that can fail.
- **Tool Function:** Describe the tool's purpose—the key question it tries to answer.
- **Potential Failure Mode:** Describe the exact nature of the potential failure of a given task.
- **Potential Failure Effects:** Define the specific impact or effects that would result if the failure occurs—the potential impact.
- **Potential Causes:** Determine what level of risk the failure presents to the project and the team.
- **Current Project Management Evaluation or Control Mechanism:** Develop and document a preventive and reactive control plan to lower the possibility of risk and the level of risk if it occurs.
- **Revised Project Plan:** Describe and document the risk-reduced project plan at a Gate Review.

Every team should present this revised, de-risked project plan at a Gate Review before starting the next phase of work. Gatekeepers should offer guidance and resources to prevent or avoid cycle-time problems during the next phase.

The analysis stage of FMEA has three ways to quantify risk:

- **SEV** (severity)**:** Rank the severity of the effect or the impact of failure mode.
- **OCC** (occurrence)**:** Identify how often failure mode may occur.
- **DET** (detectable): Identify how detectable the failure is. Define whether and how you see it coming and how to measure

the impending failures. Define how well the failure can be measured after it has occurred.

Typically the FMEA table lists the items in descending order on a scale from 1 to 10, where 10 is high risk and 1 is low risk. The three types of quantified risk values are multiplied to yield a Risk Priority Number (RPN). The failure modes are ranked by their respective RPN. The team starts with the highest RPN, which symbolizes a call to action, to devise a preventive and reactive plan to mitigate the risk. The team continues to work its way down the list, from the highest score to the lowest, to complete its preventive and reactive plan.

An RPN ranks and prioritizes the critical path of tasks based on contribution to risk. *Every task on the critical path should have an RPN and a control plan.* This should be a nonnegotiable requirement for every Gate Review. The summary gate deliverable is the cycle-time FMEA ranks and the control plans. The gatekeepers can follow the critical path by ranked risk levels and can decide what help they can give the team to protect the forecast cycle time. This helps the team and gatekeepers act in a unified fashion by including predictability in the project plan and devising reasonable ways to deliver the goals on time.

An entire project of standard marketing work can be designed, forecast, modified, and balanced for the right cycle time to meet the project's goals (see Figure 3.5 on the next page).

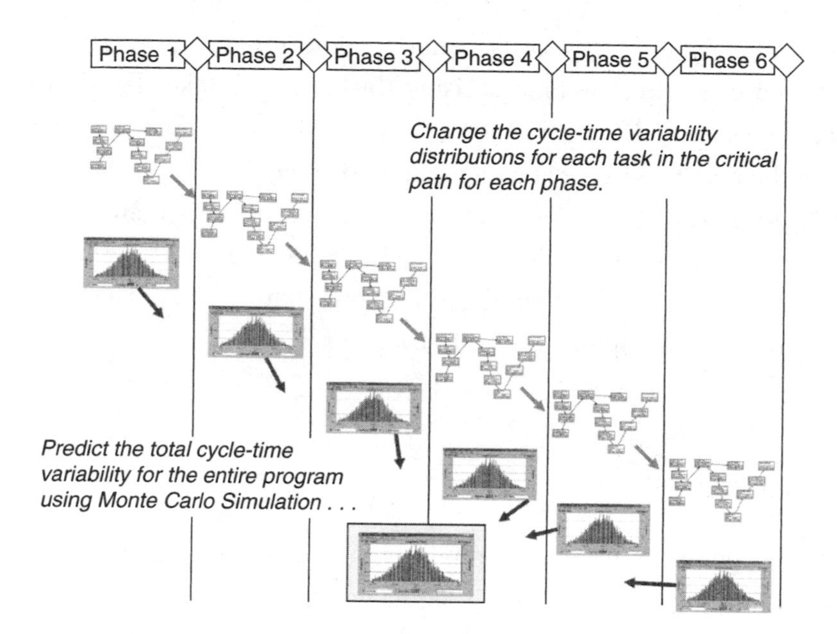

Example of Crystal Ball MC Software from Decisioneering

FIGURE 3.5 The entire project simulated.

With this kind of disciplined approach to designing cycle-time management, the team of marketing professionals can fully communicate just how long a project should take from a realistic position of tool-task flow to produce the right deliverables to fulfill the gate requirements.

Diluted core competencies on ill-defined projects with anemic deliverables—how can that possibly sustain growth? This nine-step approach to designing cycle time gives you the information necessary to enable a robust cycle-time discussion among the team, their project manager(s), and the executive sponsors. Realistic negotiations and enlightened trade-offs can take place. If you cannot see how you could possibly spend the time on such due diligence in designing and managing cycle time, we wish you luck. You will need a lot of it, because luck will be your default strategy for sustaining growth. We suggest enhancing your luck with these nine opportunities to prevent downstream cycle-time problems. Using this approach takes longer than either the bottom-up guessing or top-down deterministic approaches, but the resulting project target completion date will be more credible. The prescribed time allocated to planning may be more than what you are used to, but the results are worth it—achieving your goals the first time. Wall Street is watching, as are your shareholders. The former requires this discipline; the latter deserves it.

Summary

To this point, we have discussed the fundamental elements that transcend any well-constructed Six Sigma method. We have discussed the importance of *process thinking*, wherein processes define the road map that guides activities, and the importance of *management by fact* to inform decision-making. We have proposed three processes as the *core revenue-generating processes* in your firm: strategic, tactical, and operational. We have discussed the importance of *requirements* and how they drive the *tool-task-deliverable* linkage and *result metrics*. We have reviewed the importance of *phase-gate reviews*, *scorecards*, and *project management*. Now we are ready to explore the details of the three marketing and sales process arenas. Chapter 4 covers the strategic IDEA approach, Chapter 5, "Six Sigma in the Tactical Marketing Process," discusses the tactical UAPL structure, and Chapter 6, "Six Sigma in the Operational Marketing Process," describes the LMAD method for post-launch operations.

4

Six Sigma in the Strategic Marketing Process

Inbound Marketing for Portfolio Renewal for Products and Services

Portfolio Renewal

Growth is the new focus of the Six Sigma and Lean methods. In fact, the Six Sigma and Lean methodologies will have little influence on the future of business processes unless they help an enterprise generate sustainable growth. The word *sustainable* is important in our discussion of growth. Growth can occur by chance and can be attained by luck—but not on a sustainable basis. Processes enable sustainable performance. Processes serve as the road map that defines not only the direction in which to travel, but also the specific path, such that outcomes are repeatable and predictable. If you want to increase the probability of continued, measurable growth, value-adding tasks must be well-planned, and that, of course, requires strategic thinking and action.

Business growth has three process arenas that contribute to its sustainability: strategic, tactical, and operational (see Figure 4.1). A *flow of data* and a *system of integrated metrics* link these three areas to create consistent gains in growth. Let any one of the three arenas languish in performance, and growth will become erratic—perhaps even out of control.

FIGURE 4.1 The process triangle.

Evidence of erratic growth without established processes can be found in newly emerging industry segments. For example, today's biomedical firms are growing so fast that their revenue outpaces their ability to put key infrastructure components in place. Marketplace demands help them grow in spite of themselves. Well-managed biomedical firms, such as Genentech, Inc., know that their firm is in its infancy and are hiring Six Sigma professionals to help them build the necessary infrastructure (such as process road maps) as a stabilizing foundation for future growth. An established set of strategic-tactical-operational processes can provide a smooth transition into the next life-cycle phase.

Measuring results from marketing processes depends on the linkage and flow of data within and between them. Predictable results come from a well-planned series of tasks. Marketing tasks must be measured and controlled as an integrated system. Most companies put their money and effort into the outbound, operational marketing functions—to the exclusion of the strategic and tactical marketing functions. Worse yet, many companies simply treat marketing as a functional area, concerned with roles and responsibilities, and fail to treat marketing as a process or integrated system of measurable work. As a result, unknowing individuals often may duplicate activities or may assume that someone else has covered the bases—when in fact activities are falling between the cracks and nobody is doing them.

This chapter focuses on the strategic product planning arena called product portfolio renewal. Throughout this book, the word "product" refers to a company's generic marketplace offering and represents both a tangible product and a services offering. Regardless of the offering, the Six Sigma approach is the same and can be applied to either a tangible product or a services offering. The outcome of the four-phase strategic inbound marketing process, called IDEA (Identify-Define-Evaluate-Activate), is the definition and development of an enterprise's (or division's) portfolio renewal strategy for products and/or services.

Process Discipline in Portfolio Renewal

Tackling the job of adding process discipline and metrics to the functions within a portfolio renewal process is by far the most difficult task when integrating Six Sigma into the three marketing processes.

Adding process metrics and discipline to marketing functions in any environment is not a trivial matter. It must be done carefully, taking into account the unique culture that exists in marketing organizations. It involves both art and science. You must pay attention to how marketing teams interact with other teams within adjacent business processes—especially research and development (R&D).

This does not mean that marketing professionals get a "free pass" with regard to disciplined execution of tasks that have measurable results. It also does not mean that marketers can ignore rigorous project management to control their cycle time on projects. All teams must be accountable for delivering results, whether they are working on the future product portfolio architecture, commercializing a particular product, or supporting a product line in the post-launch environment. If you manage the critical parameters by what you do within marketing processes, you can control the results that are produced.

Can a business grow in predictable patterns if most of its focus is on post-launch line management and sales? What if it insufficiently defines and controls its tasks and deliverables within the commercialization and portfolio renewal processes? Without process discipline and clear, measurable deliverables across all three arenas (strategic, tactical, and operational), a business is unlikely to sustain growth. It is reasonable to expect inconsistent growth will result when marketing tasks and tools are underdefined and lack an integrated system of metrics linked to requirements and deliverables.

Design Process Discipline During Inbound Marketing

Many companies lack measurable process rigor in their portfolio renewal process. Time to market becomes an even tougher challenge for these firms. For example, in the automotive industry, ". . . it used to take five years to design a new-model car; today it takes two. Competing by customizing features and functions to the preferences of customers in smaller market niches is another fact of life," state Clayton Christensen and Michael Raynor in their book *The Innovator's Solution*. It is a tough environment in which to get alignment across organizational boundaries. This is particularly true of inbound marketing and R&D. A lot is at stake when structuring the company's future in the form of its product and technology portfolio architectures. If the respective portfolio architectures from marketing and R&D are incongruent, it is very hard to sustain growth. Oddly enough, the place where the largest risk lies, for developing strategic growth, is the very place where process discipline and organizational collaboration contain the least linkage, rigor, and metrics. Most notably, companies take on too many projects in the hopes that enough of them will deliver financial results that will meet the growth requirements from the business strategy. They overload the workforce—particularly understaffed marketing organizations. Applying Six Sigma thinking can ameliorate this dilemma.

The other issue is the cycle-time disruption due to technology transfer being out of sync with the activation of product commercialization projects. Typically, the technology needed to enable a new product gets developed alongside the new product. The technology is frequently immature and debilitates the product delivery schedule because the technology has to be redone. Executives expect outstanding results from project teams that cannot possibly complete all the tasks that should be done to meet their expectations. The work of executing on the new portfolio just cannot happen correctly, let alone on time.

Business leaders need to design strong, strategic alignment between product and technology portfolio development for the sake of downstream cycle-time efficiency and control. Too many simultaneously running projects, some with technology time bombs ticking away inside them, will ultimately lead to a natural outcome—unsustainable execution and growth.

The strategic product portfolio renewal process is the first of two inbound marketing and R&D/design engineering processes that we will discuss. Inbound marketing is used to characterize the flow of the marketing functions, data, and metrics for those who conduct portfolio design, renewal activities, and commercialization tasks. Engineering has two "inbound" processes that are very similar to inbound marketing. Research and technology development is the strategic component, and product design engineering is the tactical component. Both can be considered inbound engineering.

Figure 4.2 illustrates the inbound macro tasks that define what must be done during portfolio renewal.

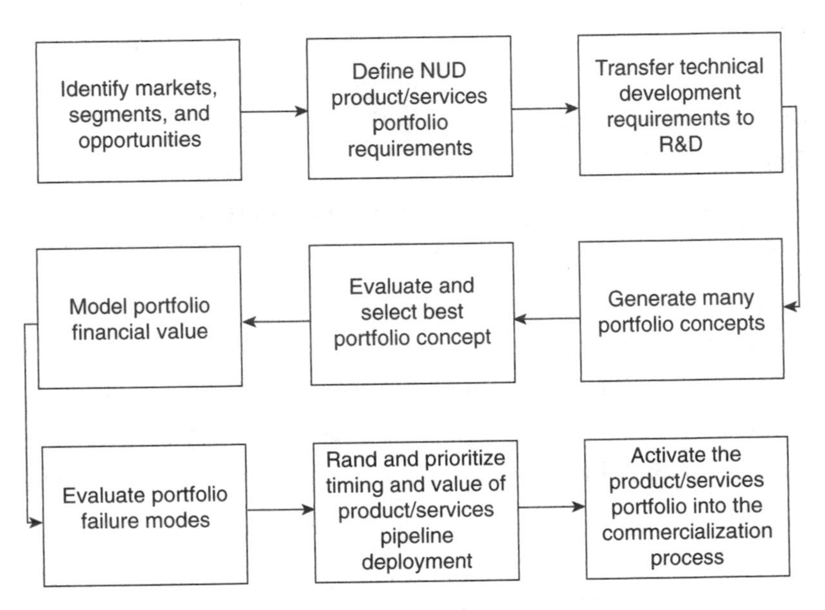

FIGURE 4.2 Portfolio renewal macro tasks.

Stepping back to develop a better context, Figure 4.3 illustrates the big picture of integrated marketing and technical functions that reside within the inbound and outbound marketing arenas.

Marketing and technical processes and functions must be linked for Six Sigma in marketing, technology development, and design to enable growth. Integrated, multifunctional teams from inbound marketing, R&D, and design and production/service support engineering

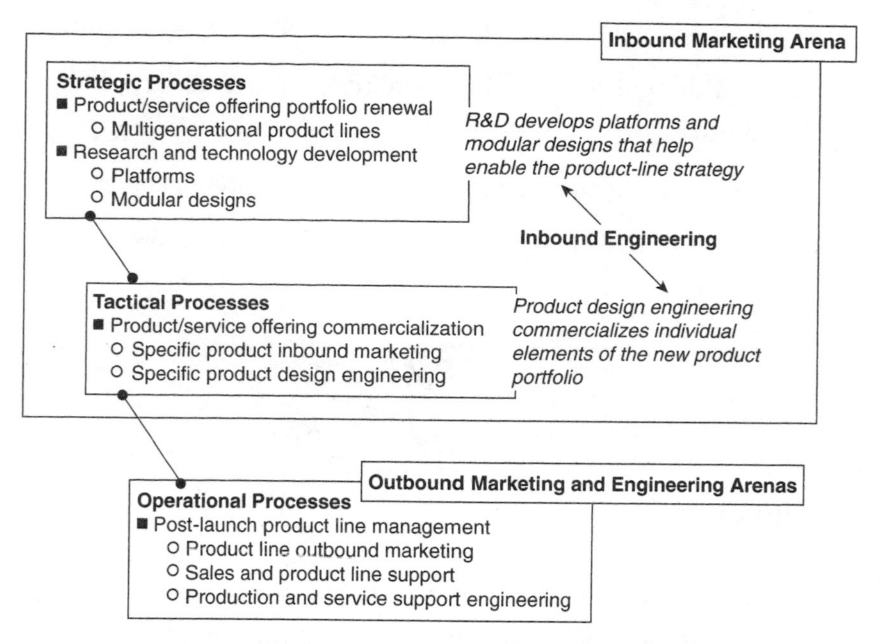

FIGURE 4.3 Integrated marketing and technical functions.

must be used across all three functional groups to develop and share data to manage risk and make decisions. Recognize that the processes to sustain growth are far more complex and require broader tool sets than the simple DMAIC problem-solving steps can handle. DMAIC can solve problems encountered along the way, but it is not *the* way. Properly designed phases and gates manage risk and make key decisions across these process arenas. The form of Six Sigma we are discussing aligns tools to tasks that you conduct on a *day-to-day basis* within a process. This new approach contrasts with DMAIC, which often is used to improve or develop on an as-needed basis, rather than a steady-state usage.

Phases of Portfolio Renewal

In the strategic marketing process environment, the four-phase IDEA process approach (see Figure 4.4) is designed to yield the definition and development of a portfolio renewal strategy:

1. **Identify** markets, their segments, and their opportunities.
2. **Define** portfolio requirements and portfolio architecture alternatives.
3. **Evaluate** portfolio alternatives against competitive portfolios.
4. **Activate** ranked and resourced individual commercialization projects.

FIGURE 4.4 The IDEA method.

A *phase* is a *period of time in which you work to produce specific results that meet the requirements* for a given project. This book takes the view that every cycle of strategic portfolio renewal is a project with distinct phases and gates. A *gate* is a *stopping point at which you review results against requirements* for a bounded set of tasks; in this case, you develop and renew the portfolio. A phase is normally designed to limit the amount of risk that can accrue before a gatekeeping team assesses the summary data that characterizes the risk of going forward. A *system of phases and gates* is how the portfolio renewal and commercialization *processes are put under control*. They define the *control plan* for the *inbound marketing processes* (see Figure 4.5).

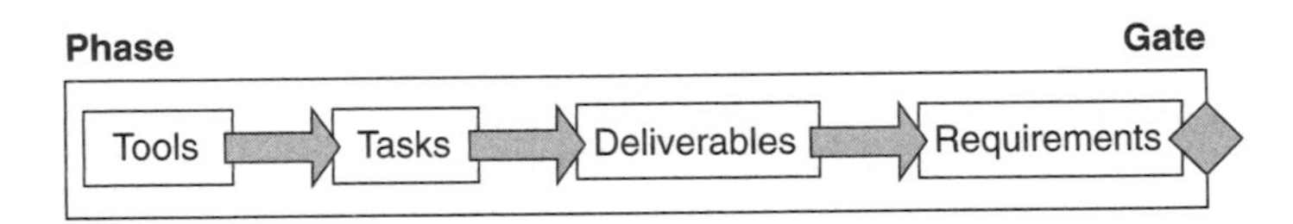

FIGURE 4.5 The generic phase-gate system.

Phases are best described by how they integrate tools, tasks, and deliverables to meet gate requirements. Gatekeeping teams establish gate requirements that in turn define exactly what tools, tasks, and deliverables are appropriate for the project. These flows are designed to define the critical marketing parameters for the inbound marketing team. The focus on requirements encourages the project team to work with the end in mind.

Identify Phase of Portfolio Renewal

The first phase in the IDEA process is heavily enabled by inbound marketing tasks and tools. It is also enabled by the inbound technology road mapping and benchmarking tasks and tools. Our focus is on the marketing component within this phase.

Before you begin the Identify phase, the business strategy must be clearly defined and documented. The business strategy breaks down into several general areas: financial growth goals, core competencies and capabilities, and innovation strategy (including marketing and technology). These goals and capabilities serve as critical criteria to evaluate, prioritize, and select potential opportunities.

Strategic inbound marketing for portfolio renewal without these elements clearly documented is like a boat without a rudder. Three main criteria determine if the Identify phase deliverables are completed:

- What is the general market?
- What are the specific segments?
- What are the specific opportunities?

To identify the target market or markets in which you intend to invest and win, the profit and growth potential is a key ingredient needed to justify the investment. Trends within a three- to five-year period will be important, but so will the key market events affecting this market, such as technology changes and new competitive entrants. When identifying the specific segments within the general markets, you must identify their unique differentiated characteristics and needs. Finally, you need to identify the specific opportunities within and across the segments. These opportunities are defined as the market's financial potential linked to unfulfilled needs (underserved segments).

Opportunities can be further broken down into two main categories, both of which contain an internal and external evaluation component. The first is financial opportunities. Relative to the potential competitors in the markets and segments, identify the financial opportunities as defined by market dynamics. The financial potential then links to specific unfulfilled marketplace needs. Next, compare this potential to your company's internal financial growth goals to evaluate whether the opportunities fulfill your objectives. The second category examines the technology and service opportunities. Identify what technology and services are currently available in the market and segments, as well as the near-term emerging capabilities that industry analysts may reveal. Then define what your company can deliver. Specifically, define the ideas embodied by the management of the critical parameters for technology and service dynamics required by the market segments. This capability must align with the innovation strategy and the company's strategic core competency.

The technology and services opportunities can be expressed as ideas that will later be gathered, documented, and converted into product, product line, and services concepts. The financial opportunities coupled with the unmet market/segment needs can be converted into customer requirements, which in aggregate can represent the new product/services offering portfolio requirements. With ideas and offering portfolio requirements in hand, you can conduct product/services offering portfolio architecting—a major goal of the Define phase for the IDEA process (see Figure 4.6).

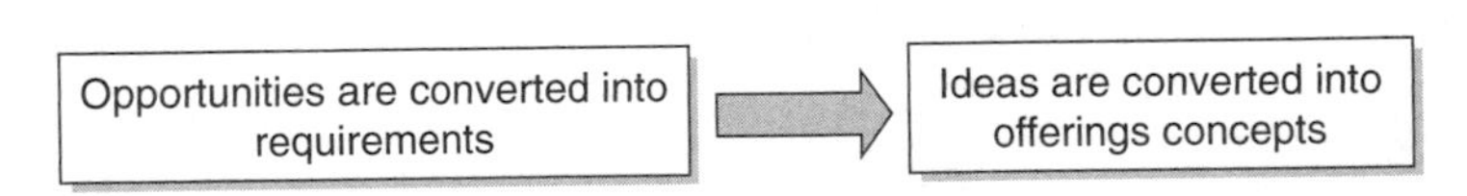

FIGURE 4.6 Requirements-ideas linkage.

The portfolio renewal process is concerned with identifying, defining, and evaluating external opportunities linked to a business's own ideas. At 3M, the business leaders use Six Sigma strategically to double the number of external opportunities linked to internal ideas before commercialization. Having implemented both Design for Six Sigma and Six Sigma for Marketing approaches, 3M now targets products that have the potential to yield three times the return compared to their previous launches (before using Six Sigma).

The final phase of the portfolio renewal process provides the linkage to the next process (commercialization). Management prioritizes and selects portfolio requirements and then sponsors, funds, and activates commercialization projects. Activating commercialization projects, in turn, converts opportunities into specific product/services offering requirements during the first phase of commercialization and then converts ideas into specific product/services offering concepts. *Portfolio requirements* are different from *specific offering requirements*. Portfolio renewal focuses on defining *portfolio requirements*. Tactical commercialization teams develop product/services requirements after a specific commercialization project has been activated from the portfolio. Frequently one hears of a specific product being done as a Design for Six Sigma (DFSS) project. However, DFSS product commercialization projects should be activated from the ranked and prioritized portfolio of projects within the portfolio renewal process.

We cannot overemphasize the importance of understanding phase requirements. They serve as a beacon—the guiding light determining the path forward. Recall that the three Identify phase requirements define the criteria against which to measure success:

- What is the general market?
- What are the specific segments?
- What are the specific opportunities?

Now you are ready to define the approach of how to do it—what deliverables fulfill the requirements, what activities are needed to produce those deliverables, and what tools are available to assist you. Remember it is *what you do and what you measure* that are important (see Figure 4.7).

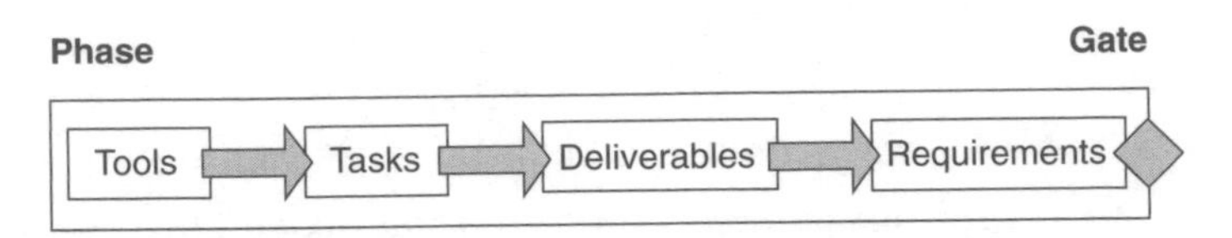

FIGURE 4.7 The tool-task-deliverable-requirement approach.

The tool-task-deliverable approach is a common structure used throughout Six Sigma. The sequence represents a cumulative effect, wherein the tool supports the task, which leads to the production of a deliverable to fulfill a specific requirement. The preceding element supports its successor.

As we detail the unique IDEA inbound marketing method to renew an offering portfolio, the following tools sections simply align the tools within a given phase. The suggested tools draw from the complete set of Six Sigma/Lean tools readily available today; no "new" tools are introduced. Because this book is an executive overview, individual tool descriptions and guidelines on how to use them fall outside the scope of this book. However, each Deliverables, Tasks, and Tools section includes a sample scorecard to show the hierarchical linkage of the tools-tasks-deliverables-requirements combination.

Identify Phase Tools, Tasks, and Deliverables

Producing the Identify phase deliverables requires an investment in numerous detailed inbound marketing tasks enabled by specific tools, methods, and best practices. Their integration ensures that the right data is being developed and summarized to fulfill all three major requirements at the Identify gate review. Figure 4.8 summarizes the tools-tasks-deliverables in the Identify phase of the IDEA process:

Tools

- Market identification and segmentation analysis
- VOC gathering methods
- Competitive benchmarking and best-practice analysis
- SWOT Analysis method
- Market perceived quality profile method
- Porter's 5 Forces Analysis method
- Market behavioral dynamics map methods
- IDEA capture and database development tools
- Project management tools

Tasks

- Define business strategy and financial goals
- Define innovation strategy
- Define markets
- Define market segments
- Define opportunities across markets and within segments
- Gather and translate VOC data

Deliverables

- Documented growth goals
- Documented core competencies
- Documented innovation strategy
- Documented markets
- Documented market segments
- Documented opportunities

Supporting Deliverables

- Market and segment behavioral dynamics map
- Competitive benchmarking data and trend charts
- Market perceived quality profile and gap matrix
- Porter's 5 Forces chart
- SWOT matrix

FIGURE 4.8 The tools-tasks-deliverables in the Identify phase of the IDEA process.

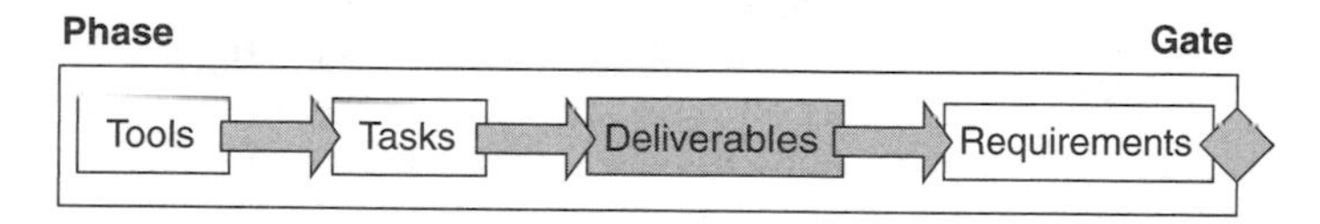

The deliverables at the Identify gate fall into two categories—major deliverables and supporting deliverables. Both are critical outputs for this phase and should be inspected for completeness and accuracy. The six major deliverables are as follows:

- **Documented growth goals**, including financial targets and tolerances.
- **Documented core competencies**, identifying availability and readiness.
- **Documented innovation strategy**, including the marketing strategy component.
- **Documented markets**, defining the macro profit pool for each market.
- **Documented market segments**, describing the micro profit pools within each.

- **Documented opportunities**, describing the financial potential and customer need dynamics.

The five supporting Identify gate deliverables are in-process deliverables that serve as inputs to the major deliverables. The supporting deliverables are as follows:

- **Market and segment behavioral dynamics map**: This can be depicted in three formats: customer behavioral process maps, value chain diagrams, and key event time lines.
- **Competitive benchmarking data and trend analysis**
- **Market perceived quality profile and gap analysis**
- **Porter's 5 Forces Analysis and segmented risk profiles**
- **SWOT matrix**: Some Six Sigma practitioners prefer analyzing "challenges" rather than "weakness"; hence, the SCOT analysis matrix may be substituted for SWOT. Both SWOT and SCOT analyses match the company's resources and its capabilities to the competitive environment within a market segment.

The scorecard, as shown in Table 4.1, is used by marketing gatekeepers who manage risk and make functional gate decisions for a specific project as part of the portfolio of projects being conducted by the marketing organization.

Columns 1 and 6 align the gate deliverable to a gate requirement. Each deliverable is justified as it contributes to meeting a gate requirement. (Never produce a deliverable if you can't justify its ability to fulfill a gate requirement.)

Grand Average Tool Score (GATS) illustrates aggregated tool quantification across the three scoring dimensions. (A high GATS indicates that a group of tasks is underwriting a gate deliverable.)

% Task Completion is scored on a 10 to 100% scale. This measure is critical if you want to understand how completely a group of related tasks are fulfilling a major gate deliverable.

Color coding is intended to illustrate the nature of the risk accrual for each major deliverable within this phase of the process.

TABLE 4.1 Sample Identify Gate Deliverables Scorecard

1	2	3	4	5	6
SSFM Gate Deliverable	**Grand Average Tool Score**	**% Task Completion**	**Task Results Versus Gate Deliverable Requirement**	**Risk Color (R, Y, G)**	**Gate Requirement(s) (General Phase Requirement)**
Documented growth goals					Define current organizational growth goals and related competencies to achieve goals (current state)
Documented core competencies					
Documented markets					Identify target markets (general market)
Documented market segments					Identify target market segments (specific market)
Documented opportunities					Identify external opportunities based on current data (specific opportunity)
Documented VOC					
Documented new, unique, and difficult (NUD) customer needs					

(*continues*)

TABLE 4.1 Sample Identify Gate Deliverables Scorecard (continued)

1	2	3	4	5	6
SSFM Gate Deliverable	**Grand Average Tool Score**	**% Task Completion**	**Task Results Versus Gate Deliverable Requirement**	**Risk Color (R, Y, G)**	**Gate Requirement(s) (General Phase Requirement)**
Documented innovation ideas					Generate and document internal innovation ideas in database (specific opportunity)
Documented define phase project plan					Document next phase's project plans and risk assessment at Identify gate
Market and segment behavioral dynamics map					Identify trends in market and segment behavioral dynamics (general and specific market)
Competitive benchmark data and trend analysis					Document current competitive behaviors and trends (specific opportunity)
Estimated market perceived quality profile and gap analysis					Estimate market perceived quality gaps (specific opportunity)
Porter's 5 Forces Analysis and segmented risk profiles					Identify risks and opportunities across markets and segments (specific opportunity)
SWOT matrix					

Color code risk definitions are found in Chapter 2, "Measuring Marketing Performance and Risk Accrual Using Scorecards."

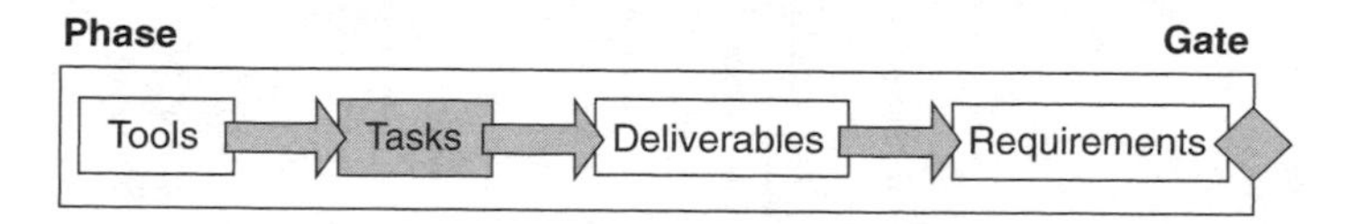

The Identify phase necessitates many tasks that superficially seem overwhelming. However, several can be done in parallel or naturally follow their predecessors. To ensure completeness, the task list defines the obvious activities needed to produce this phase's deliverables. Here are the tasks to be performed within the Identify phase that produce the deliverables:

- Define the business strategy and financial goals.
- Define the innovation strategy.
- Define the markets.
- Define the market segments.
- Define opportunities across markets and within segments.
- Gather and translate "over-the-horizon" VOC data.
- Document NUD customer needs across segments.
- Construct market and segment behavioral dynamics maps.
- Conduct competitive benchmarking (for marketing, sales channel, and technical disciplines).
- Create a database of internal ideas based on opportunity categories.
- Create a Define phase project plan and risk analysis.

Although the project management activities may be supported by another discipline (such as the technical community), marketing often owns the overall management of this strategic project and is accountable for (at minimum) marketing's deliverables and overall risk mitigation planning.

TABLE 4.2 Sample Identify Phase Task Scorecard

1	2	3	4	5	6
SSFM Task	**Average Tool Score**	**% Task Completion**	**Task Results Versus Gate Requirement**	**Risk Color (R, Y, G)**	**Gate Deliverable(s)**
Interview stakeholders and understand goals, competencies, and boundary conditions					Documented growth goals
					Documented core competencies
Define markets					Documented markets
Define market segments					Documented segments
Define opportunities across markets and within segments					Documented opportunities (estimated market perceived quality profiles and gap analyses, Porter's 5 Forces Analysis and segmented risk profiles, SWO)

Gather and translate "over-the-horizon" VOC data					Documented VOC
Document NUD customer needs across segments					Documented NUD customer needs
Create database of internal ideas based on opportunity categories					Documented innovation ideas
Create Define phase project plan and risk analysis					Documented Define phase project plan
Construct market and segment behavioral dynamics maps					Behavioral dynamics maps
Conduct competitive benchmarking (for marketing, sales channels, and technical)					Competitive benchmarking data and trend analysis

The scorecard, as shown in Table 4.2, is used by marketing project team leaders who manage major tasks and their time lines as part of their project management responsibilities.

Columns 1 and 6 align the task to a specific gate deliverable for justifying the task. (Never conduct a task if you can't justify its ability to produce a gate deliverable and fulfill a gate requirement.)

Average Tool Score (ATS) illustrates overall tool quantification across the three scoring dimensions. (A high ATS indicates that a task is being underwritten by proper tool usage.)

% Task Completion is scored on a 10 to 100% scale. This measure is critical if you want to understand how completely each specific, major task is being done.

Color coding is intended to illustrate the nature of the risk accrual for each major task within this phase of the process. Color code risk definitions are found in Chapter 2.

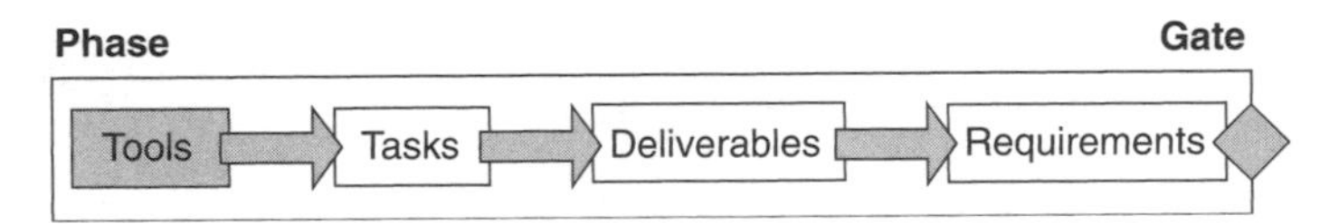

The tools available for the Identify phase also encompass methods and best practices. These tools enable the tasks associated with fulfilling the Identify deliverables and span the following items:

- **Market identification and segmentation analysis tools**: Three different types of tools can be used in this phase: secondary market research and data-gathering tools; economic and market trend forecasting tools; and statistic (multivariate) data analysis for cluster, factor, and discriminate analysis.
- **VOC gathering methods**: VOC methods cover the primary market research and data-gathering tools.
- **Market and segment VOC data processing best practices**: This tool set includes the KJ Diagramming Method (developed by Japanese anthropologist Jiro Kawakita [KJ]), plus questionnaire and survey design methods.

- **SWOT Analysis method**: Strengths (S) or weaknesses (W) external to the firm can be classified as opportunities (O) or threats (T).
- **Market perceived quality profile method**
- **Porter's 5 Forces Analysis method**: A model to analyze the five competitive forces: suppliers, customers, new entrants, substitutes, and competition.
- **Market behavioral dynamics map methods**
- **Competitive benchmarking tools**
- **IDEA capture and database development tools**
- **Project management tools**: The Six Sigma-based tools that support project management include Monte Carlo Simulation, a statistical cycle-time design and forecasting tool, and Failure Modes & Effects Analysis (FMEA) for cycle-time risk assessment.

The scorecard, as shown in Table 4.3, is used by marketing teams who apply tools to complete tasks.

Columns 1 and 7 align the tool to a specific task and its requirement for justifying the use of the tool. (Never use a tool if you can't justify its ability to fulfill a task.)

Quality of Tool Usage, Data Integrity, and Results Versus Requirements are scored on a scale from 1 to 10 using the simple scoring templates provided in Chapter 2.

Average Tool Score illustrates overall tool quantification across the three scoring dimensions. (A high ATS indicates that a task is being underwritten by proper tool usage.)

Data Summary is intended to illustrate the nature of the data and the specific units of measure (attribute or continuous data, scale or specific units of measure).

Control, risk management, and key decision-making are greatly enhanced when you clearly define the requirements, deliverables,

TABLE 4.3 Sample Identify Phase Tools Scorecard

1	2	3	4	5	6	7
SSFM Tool	**Quality of Tool Usage**	**Data Integrity**	**Results Versus Requirements**	**Average Tool Score**	**Data Summary, Including Type and Units**	**Task**
Interview guide						Interview stakeholders and understand goals, competencies, and boundary conditions
Market identification and segmentation analysis tools						Define markets and segments
Estimated market perceived quality profile gap analysis method						Define opportunities across markets and within segments
Porter's 5 Forces Analysis method						
SWOT Analysis method						

VOC gathering tools and methods						Gather and translate "over-the-horizon" VOC data
Market and segment VOC data processing best practices						Document NUD customer needs across segments
Concept-generation tools, TRIZ, brainstorming, mind mapping						Create database of internal ideas based on opportunity categories
Project management tools, Monte Carlo, FMEA, RACI Matrix						Create Define phase project plan and risk analysis
Market behavioral dynamics mapping methods						Construct market and segment behavioral dynamics maps
Competitive benchmarking tools						Conduct competitive benchmarking and define opportunities across and within segments

tasks, and enabling tools within the Identify phase of portfolio renewal (see Figure 4.9). If these four layers are ambiguous, detailed root causes of poor sustainability in growth begin to emerge. When these four elements begin to degrade, you have begun down the slippery slope to poor performance that will eventually result in erratic growth.

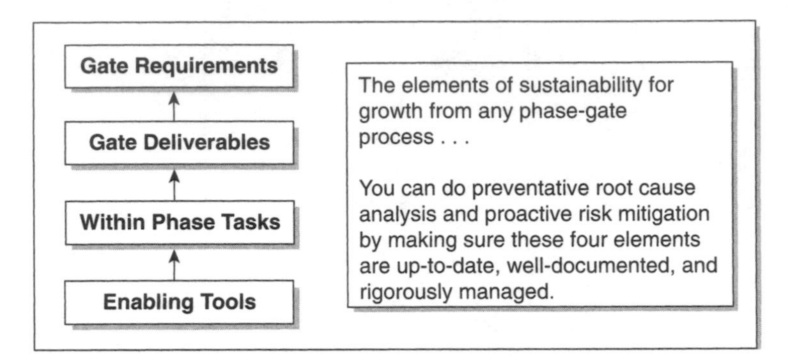

FIGURE 4.9 Elements of the phase-gate process.

The Six Sigma discipline is commonly associated with the design and application of a control plan once a process is established and known to be capable. This is a control plan for conducting strategic marketing functions in the portfolio renewal process. We often hear of control plans for manufacturing processes. This is the marketing equivalent when conducting work in a phase-gate process environment. It applies across strategic, tactical, and operational marketing and sales processes.

Define Phase Tools, Tasks, and Deliverables

Producing the Define phase deliverables requires an investment in numerous detailed inbound marketing tasks enabled by specific tools, methods, and best practices. Their integration ensures that the right data is being developed and summarized to fulfill the major requirements at the Define gate review.

The phase requirements for the Define gate are threefold:

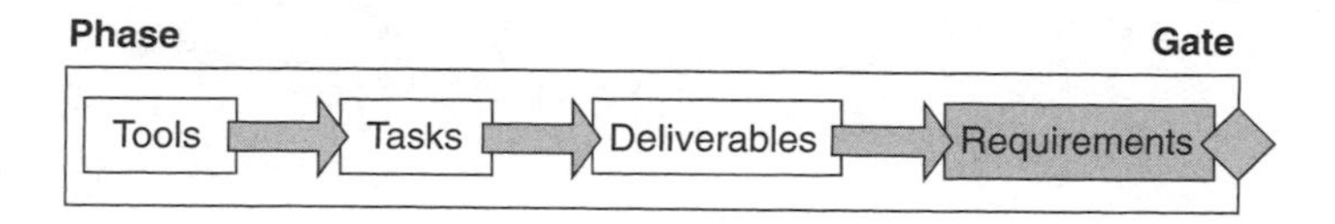

- **Documented NUD portfolio requirements**. The new requirements (comprehending customer targets and ranges) describe what no one in competitive or adjacent industries currently fulfills. The unique requirements (again consisting of targets and ranges) specify that which your competitors or adjacent industries currently satisfy—but that you currently do not. The difficult requirements are explicit opportunities to address if you can use your core competencies to overcome associated hurdles.
- **Documented requirements by segments** that differentiate that which is common across the targeted segments and that which is different or unique across and within the segments.
- **Identified candidate portfolio architectures** entail various mixes of product/service offering ideas. A variety of opportunities are matched with your various ideas. Several different architectures are generated to fulfill the portfolio requirements across, and within, the markets and their segments (see Figure 4.10).

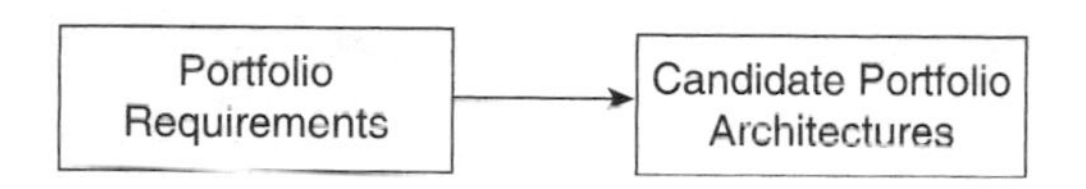

FIGURE 4.10 Requirements-candidate linkage.

Figure 4.11 shows the flow of knowledge into the Define phase.

First, you define markets that have sufficient profit potential to fulfill your business strategy. These markets possess New, Unique, and Difficult needs that align with your core competencies (see Figure 4.12).

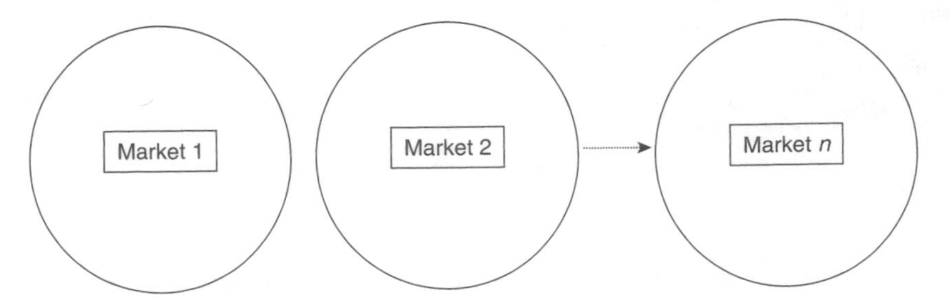

FIGURE 4.11 The flow of knowledge.

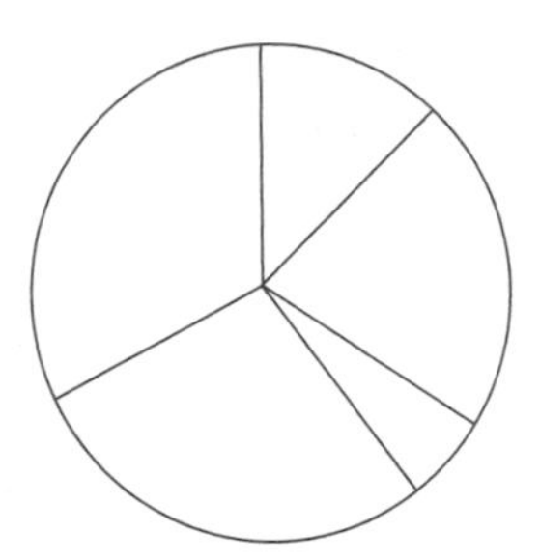

FIGURE 4.12 The NUD circle.

Next, with statistically significant data, determine that you have defined differentiated needs between the market segments within the general markets. Ensure that you have NUD needs ready to translate into a set of portfolio requirements that possess common and differentiated targets and fulfillment ranges. It is these requirements that drive the portfolio architecting process. Innovation at this level is the most strategic form of creativity and ideation a company can conduct. If you fail to innovate here, the sustainability of growth is at risk.

Matching ideas to opportunities within and across segments helps ensure that you are generating the right mix of alternatives for your portfolio architecting process (see Figure 4.13). Blending and considering alternatives between your ideas and market opportunities helps lower the risk of launching an unbalanced portfolio. This helps stabilize growth potential on a systematic basis.

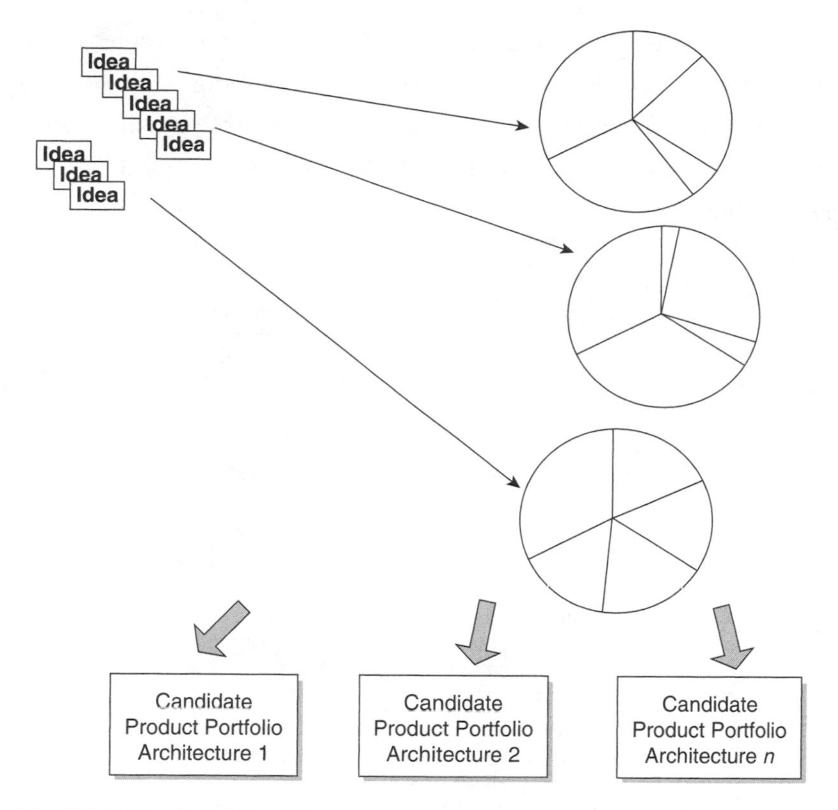

FIGURE 4.13 Matching ideas to opportunities.

The Define phase work integrates and documents mixes of ideas that are aligned with the real opportunities that exist within and across your markets and segments. Innovative ideas are essential to exploiting market opportunities. Constant creating, gathering, prioritizing, and pruning of ideas are part of the dynamics that make companies grow. If there is a lull in this activity, you can easily see how a dip in growth can occur. Let your ideation die out for a period of time, and you can see a direct cause-and-effect relationship relative to sustaining growth. Ideas can come from anywhere, and constant incubation transcends the IDEA phases. Harvesting and integrating ideas into candidate portfolio architectures in the Define phase is a must for the IDEA process to work properly.

To summarize at this point, notice how the marketplace and customer needs drive ideation. Ideas can be linked and matched to opportunities for portfolio architecting. Portfolio requirements and alternative architectures drive technology planning and development. Figure 4.14 shows how a product renewal strategy may evolve. (Note that a services renewal could follow a similar and perhaps parallel flow.)

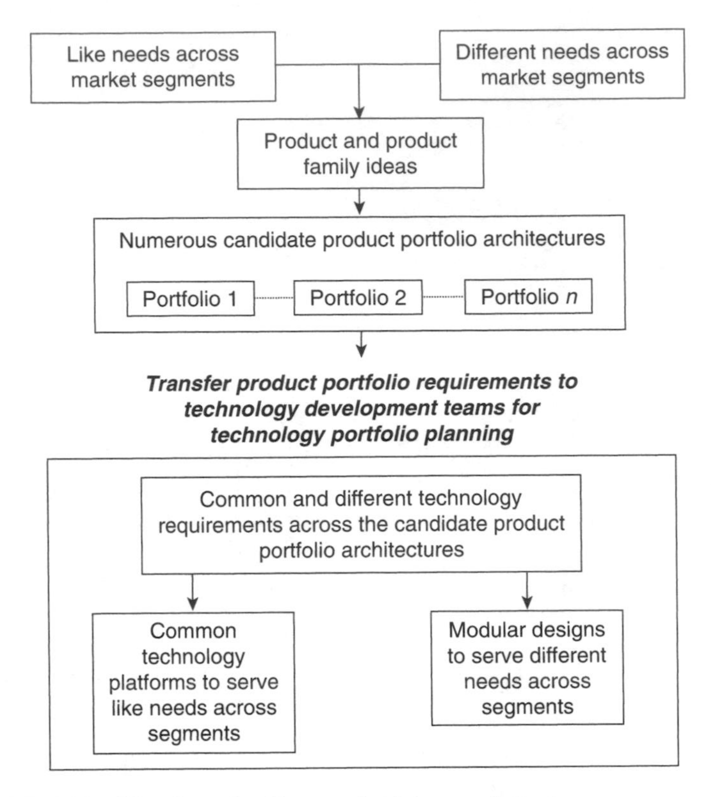

FIGURE 4.14 Idea flow for the product renewal strategy.

The portfolio renewal team defines the requirements for the new product (or services offering) portfolio architecture based on NUD needs from the segments. Numerous ideas are matched with market opportunities that bound the target environment for the new portfolio. The ideas are then grouped and linked to form candidate portfolio architectures. Each of the architectures represents one mix of offering ideas that has the potential to fulfill the business strategy and meet financial growth goals. Once a number of candidate portfolio architectures are defined, they are transferred to the teams that conduct technology planning and development, along with the portfolio requirements. The early work of technology development is now linked to the output from the IDEA process. Technology is no longer allowed to develop in a vacuum. It is tied to your strategy in a very traceable and measurable format.

Figure 4.15 lists the required tools-tasks-deliverables for the Define phase.

Tools

- Statistical survey design and analysis
- Statistical data analysis
- QFD tool
- Product portfolio architecting methods
- Project management tools

Tasks

- Design and conduct VOC validation surveys
- Analyze VOC
- Translate VOC into portfolio requirements
- Refine market segmentation
- Define candidate portfolio architectures
- Refine market and segment behavioral dynamics map
- Create EVALUATE phase project plan and risk analysis

Deliverables

- Documented NUD portfolio requirements
- Documented requirements by segments
- Identified candidate product/services offering portfolio architectures

Supporting Deliverables

- Segmentation statistics summary
- VOC-based requirements data
- Common requirements across segments
- Differentiated requirements across segments
- Customer survey results

SSFM Portfolio Renewal
I D E A

FIGURE 4.15 The tools-tasks-deliverables in the Define phase of the IDEA process.

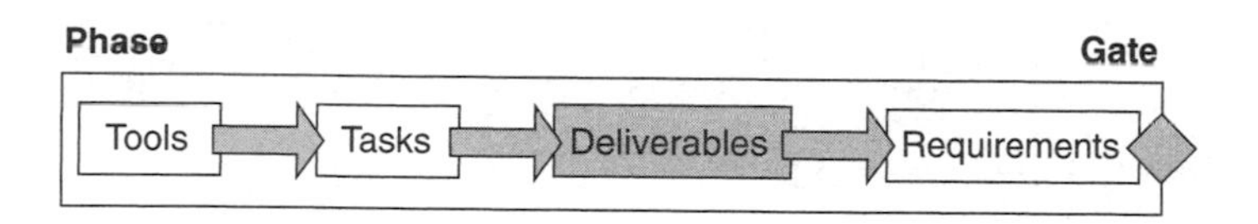

As with Identify, the Define phase has both major and supporting deliverables to be completed to exit this phase and continue to

Evaluate. The Define gate major deliverables closely follow the phases' requirements and need little introduction: documented NUD portfolio requirements, documented requirements by segments, and documented candidate product and/or services portfolio architectures. The five Define supporting deliverables are as follows:

- **Segmentation statistics summary,** which includes cluster, factor, and discriminate analysis.
- **VOC-based requirements,** covering both target and range data.
- **Common customer requirements** across segments.
- **Differentiated customer requirements** across and within segments.
- **Customer survey** and validation results.

The scorecard, as shown in Table 4.4, is used by marketing gatekeepers who manage risk and make functional gate decisions for a specific project as part of the portfolio of projects being conducted by the marketing organization.

Columns 1 and 6 align the gate deliverable to a gate requirement. Each deliverable is justified as it contributes to meeting a gate requirement. (Never produce a deliverable if you can't justify its ability to fulfill a gate requirement.)

Grand Average Tool Score (GATS) illustrates aggregated tool quantification across the three scoring dimensions. (A high GATS indicates that a group of tasks is underwriting a gate deliverable.)

% Task Completion is scored on a 10 to 100% scale. This measure is critical if you want to understand how completely a group of related tasks are fulfilling a major gate deliverable.

Color coding is intended to illustrate the nature of the risk accrual for each major deliverable within this phase of the process. Color code risk definitions are found in Chapter 2.

TABLE 4.4 Sample Define Gate Deliverables Scorecard

1	2	3	4	5	6
SSFM Gate Deliverable	**Grand Average Tool Score**	**% Task Completion**	**Tasks Results Versus Gate Deliverable Requirement**	**Risk Color (R, Y, G)**	**Gate Requirement(s) (General Phase Requirement)**
Segmentation VOC statistics summary documents					Identify NUD portfolio requirements across and within segment (NUD portfolio requirement)
VOC-based requirements documents					
Documented summary results from customer surveys					
Documented common customer requirements across and within segments					
Documented summary NUD portfolio requirements					
Documented differentiated customer requirements across and within segments					Document differentiated NUD requirements across segments for modularity purposes (documented requirements by segments)

(continues)

TABLE 4.4 Sample Define Gate Deliverables Scorecard (continued)

1	2	3	4	5	6
SSFM Gate Deliverable	**Grand Average Tool Score**	**% Task Completion**	**Tasks Results Versus Gate Deliverable Requirement**	**Risk Color (R, Y, G)**	**Gate Requirement(s) (General Phase Requirement)**
VOC-based portfolio requirements documents translated into technical requirements; houses of quality by segment					Translate portfolio requirements into technical requirements (identified candidate portfolio architectures)
Documented product/service portfolio architectures					Create and document several identified candidate portfolio architectures
Updated market and segment behavioral dynamics maps					Show how established market and segment behavior mitigate risk and substantiate viability of concept
Documented Evaluate project plans and risk analysis					Document Evaluate phase project plans and risk assessment

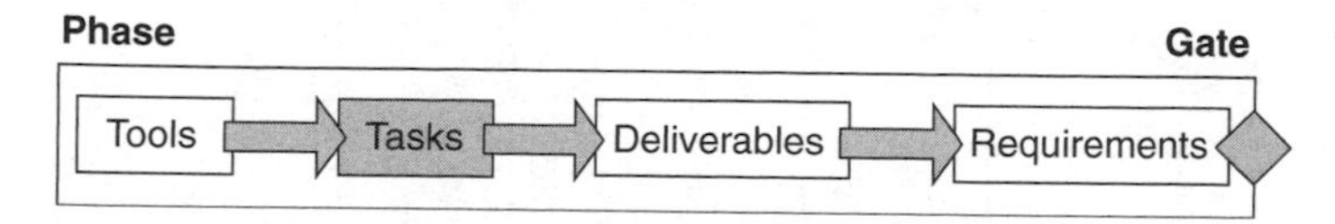

The tasks to be performed within the Define phase that produce the Define gate deliverables are as follows:

1. Design and conduct VOC validation surveys to find common versus differentiated requirements for use in time and circumstance (containing the targets and ranges).
2. Conduct statistical analysis on VOC survey data sets to refine segment identification.
3. Translate NUD customer needs into portfolio requirements (with translation into business vocabulary with measurable units).
4. Refine market segments based on VOC validation survey results.
5. Define candidate offering portfolio architectures built from mixes of product and/or services offering ideas that meet portfolio requirements across and within segments.
6. Refine market and segment behavioral dynamics maps.
7. Create Evaluate phase project plan and risk analysis.

The scorecard, as shown in Table 4.5, is used by marketing project team leaders who manage major tasks and their timelines as part of their project management responsibilities.

Columns 1 and 6 align the task to a specific gate deliverable for justifying the task. (Never conduct a task if you can't justify its ability to produce a gate deliverable and fulfill a gate requirement.)

Average Tool Score (ATS) illustrates overall tool quantification across the three scoring dimensions. (A high ATS indicates that a task is being underwritten by proper tool usage.)

% Task Completion is scored on a 10 to 100% scale. This measure is critical if you want to understand how completely each specific, major task is being done.

TABLE 4.5 Sample Define Phase Task Scorecard

1	2	3	4	5	6
SSFM Task	**Average Tool Score**	**% Task Completion**	**Task Results Versus Gate Requirement**	**Risk Color (R, Y, G)**	**Gate Deliverable(s)**
Design and conduct VOC validation surveys to help define common and differentiated targets and ranges					Documented common and differentiated customer NUD requirements across and within segments
Conduct statistical analysis on VOC survey data sets to help refine segment boundaries					Segmentation VOC statistics summary documents; factor, cluster, and discriminant analysis results
Refine market segments based on VOC survey results					Segmentation VOC statistics summary documents; refined market segmentation document

Translate NUD customer needs into technical portfolio requirements					VOC-based portfolio requirements documents; Houses of Quality by segment
Define candidate portfolio architectures					Documented candidate portfolio architectures
Refine market and segment behavioral dynamics maps					Updated market and segment behavioral dynamics maps
Create Evaluate phase project plan and risk analysis					Documented Evaluate phase project plan and risk summary

Color coding is intended to illustrate the nature of the risk accrual for each major task within this phase of the process. Color code risk definitions are found in Chapter 2.

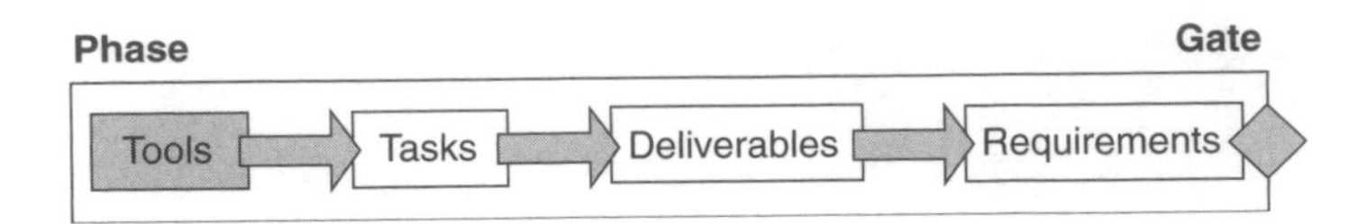

The tool set that supports the Define tasks to provide the Define deliverables includes the following software, methods, and best practices:

- **Statistical survey design and analysis**.
- **Statistical data analysis,** such as descriptive and inferential statistics, sample size determination, t-Test analysis, and multivariate analysis.
- **QFD** to translate VOC into portfolio requirements, and the portfolio Houses of Quality (HOQ).
- **Product portfolio architecting methods**, which can also be applied to services.
- **Project management tools**, the Six Sigma-based tools that support project management include Monte Carlo Simulation, a statistical cycle-time design and forecasting tool, and FMEA for cycle-time risk assessment.

The scorecard, as shown in Table 4.6, is used by marketing teams who apply tools to complete tasks.

Columns 1 and 7 align the tool to a specific task and its requirement for justifying the use of the tool. (Never use a tool if you can't justify its ability to fulfill a task.)

Quality of Tool Usage, Data Integrity, and Results Versus Requirements are scored on a scale of 1 to 10 using the simple scoring templates provided in Chapter 2.

TABLE 4.6 Sample Define Phase Tools Scorecard

1	2	3	4	5	6	7
SSFM Tool	**Quality of Tool Usage**	**Data Integrity**	**Results Versus Requirements**	**Average Tool Score**	**Data Summary, Including Type and Units**	**Task**
Statistical survey design methods						Design and conduct VOC validation surveys to help define common and differentiated targets and ranges
Descriptive, inferential, and multivariate statistical analysis tools (t-Tests, factor, discriminant, and cluster analysis)						Conduct statistical analysis on VOC survey data sets to help refine segment boundaries; refine market and segment behavioral dynamics maps
QFD/HOQ						Translate NUD customer needs into technical portfolio requirements
Product portfolio architecting methods						Define candidate portfolio architectures
Project management tools						Create Evaluate phase project plan and risk analysis

Average Tool Score illustrates overall tool quantification across the three scoring dimensions. (A high ATS indicates that a task is being underwritten by proper tool usage.)

Data Summary is intended to illustrate the nature of the data and the specific units of measure (attribute or continuous data, scale or specific units of measure).

Evaluate Phase Tools, Tasks, and Deliverables

The third phase of the IDEA process is Evaluate. Producing the Evaluate phase deliverables requires an investment in numerous detailed inbound marketing tasks enabled by specific tools, methods, and best practices. Their integration ensures that the right data is being developed and summarized to fulfill the major requirements at the Evaluate gate review.

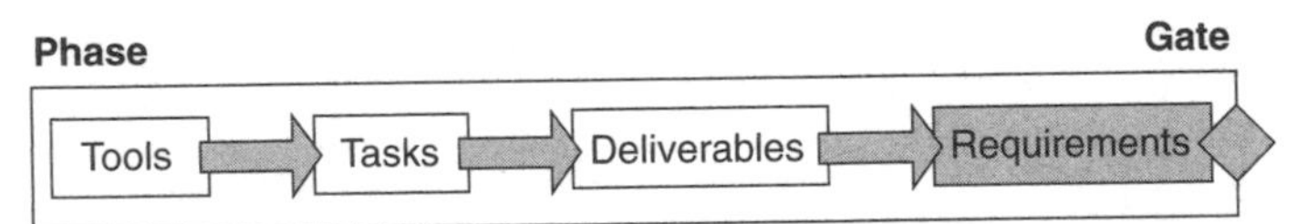

The major Evaluate gate requirements are threefold. The first requirement is a summary of the candidate product/services offering portfolio architectures. The second requirement is selecting the best portfolio architecture from among the candidates based on market potential and business case analysis. A hybrid portfolio architecture is acceptable to blend the best balance of elements from the several candidate architectures. This is called collaborative innovation. The third requirement is a financial potential and entitlement summary from the selected portfolio architecture as a comparison to the business's overall growth goals.

Figure 4.16 summarizes the Evaluate phase's tools, tasks, and deliverables.

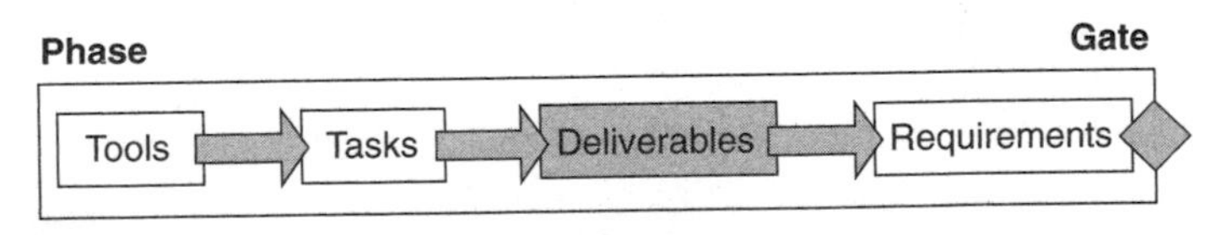

Tools

- Financial modeling and forecasting
- Business case development
- Portfolio balancing methods
- Real-Win-Worth Analysis
- Pugh method
- Project management tools

Tasks

- Create preliminary business case
- Develop portfolio evaluation criteria
- Evaluate portfolio architecture
- Assess portfolio financials
- Create ACTIVATE phase project plan and risk analysis

Deliverables

- Documented growth portfolio
- Documented portfolio architecture
- Documented risk assessment
- Documented real-win-worth analysis

Supporting Deliverables

- Preliminary business case
  - Financials
  - Value propositions
  - Market dynamics
  - Risks
  - Benchmarking

FIGURE 4.16 The tools-tasks-deliverables in the Evaluate phase of the IDEA process.

The four major deliverables at the Evaluate gate are as follows:

- **Documented growth potential** of the balanced offering portfolio architecture (including financial targets and tolerances)
- **Documented product portfolio architecture**
- **Documented risk assessment** for the portfolio that shows a balance across multiple dimensions of risk, with a portfolio FMEA to identify possible failures, respective frequencies, and potential impacts of those failures
- **Documented Real-Win-Worth (RWW) Analysis** across the portfolio's elements

The supporting Evaluate gate deliverables all center on a set of preliminary business cases for the elements in the portfolio. The preliminary business case needs to include the following:

- Financial models
- Value propositions
- Market dynamics and fit with trends
- Technical risk profiles

TABLE 4.7 Sample Evaluate Gate Deliverables Scorecard

1	2	3	4	5	6
SSFM Gate Deliverable	**Grand Average Tool Score**	**% Task Completion**	**Tasks Results Versus Gate Deliverable Requirements**	**Risk Color (R, Y, G)**	**Gate Requirement(s) (General Phase Requirement)**
Documented candidate portfolio architectures					Summarize each candidate portfolio architecture (summary candidate portfolio)
Documented best portfolio architecture					Select the best hybrid portfolio from among the candidates in light of the portfolio requirements against a best-in-class datum portfolio architecture (best portfolio candidate)
Documented financial growth potential and preliminary business cases for each candidate portfolio architecture					Forecast financial potential and entitlement summary from the selected (best) portfolio architecture to compare against the business's growth goals (financial potential)
Documented RWW Analysis for each candidate portfolio architecture					Conduct RWW Analysis on each candidate portfolio architecture
Documented risk assessment for each candidate portfolio architecture					Identify and quantify the risks associated with each candidate portfolio architecture
Documented Activate project plans and risk analysis					Have the Activate project plans and risk analysis

- Competitive benchmarking data and trend analysis
- Market perceived quality profile and gap analysis
- Porter's 5 Forces Analysis and segmented risk profiles
- SWOT Analysis and segmented risk profiles

The scorecard, as shown in Table 4.7, is used by marketing gatekeepers who manage risk and make functional gate decisions for a specific project as part of the portfolio of projects being conducted by the marketing organization.

Columns 1 and 6 align the gate deliverable to a gate requirement. Each deliverable is justified as it contributes to meeting a gate requirement. (Never produce a deliverable if you can't justify its ability to fulfill a gate requirement.)

Grand Average Tool Score (GATS) illustrates aggregated tool quantification across the three scoring dimensions. (A high GATS indicates that a group of tasks is underwriting a gate deliverable.)

% Task Completion is scored on a 10 to 100% scale. This measure is critical if you want to understand how completely a group of related tasks are fulfilling a major gate deliverable.

Color coding is intended to illustrate the nature of the risk accrual for each major deliverable within this phase of the process. Color code risk definitions are found in Chapter 2.

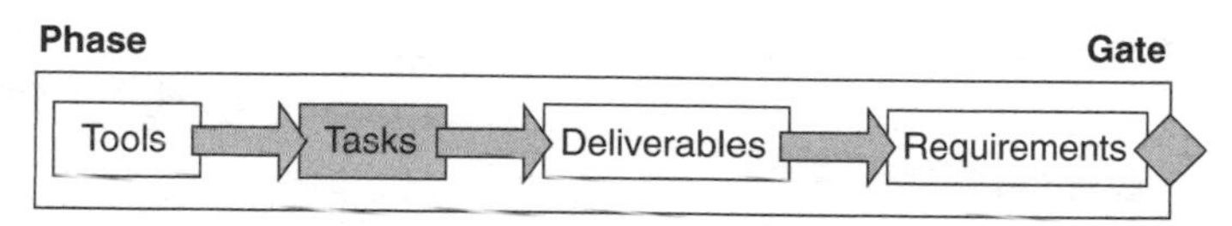

The tasks to be performed within the Evaluate phase that produce the deliverables include the following:

1. Create preliminary business cases for elements of the candidate portfolios.
2. Develop a portfolio evaluation criteria and benchmark portfolio architecture.
3. Evaluate and select the best portfolio architecture from the candidates. (It may be a hybrid of the candidates.)

TABLE 4.8 Sample Evaluate Phase Task Scorecard

1	2	3	4	5	6
SSFM Task	**Average Tool Score**	**% Task Completion**	**Task Results Versus Gate Requirements**	**Risk Color (R, Y, G)**	**Gate Deliverable(s)**
Generate candidate portfolio architectures					Documented candidate portfolio architectures
Evaluate portfolio architecture candidates and select a best portfolio architecture; conduct the Pugh Process					Documented best portfolio architecture from candidate alternatives
Create preliminary business cases for the elements of each candidate portfolio architecture					Documented financial growth potential and preliminary business cases for each candidate portfolio architecture
Conduct RWW Analysis on candidate portfolio architectures					Documented RWW Analysis for each candidate portfolio architecture
Conduct risk analysis on each candidate portfolio architecture (FMEA)					Documented risk assessment for each candidate portfolio architecture
Develop Activate project plans and project risk analysis					Documented Activate phase project plans and risk analysis

4. Assess portfolio financials, including Net Present Value (NPV), Expected Commercial Value (ECV), and Return on Investment (ROI), on selected portfolio architectures.

5. Create Activate phase project plan and risk analysis.

The scorecard, as shown in Table 4.8, is used by marketing project team leaders who manage major tasks and their time lines as part of their project management responsibilities.

Columns 1 and 6 align the task to a specific gate deliverable for justifying the task. (Never conduct a task if you can't justify its ability to produce a gate deliverable and fulfill a gate requirement.)

Average Tool Score (ATS) illustrates overall tool quantification across the three scoring dimensions. (A high ATS indicates that a task is being underwritten by proper tool usage.)

% Task Completion is scored on a 10 to 100% scale. This measure is critical if you want to understand how completely each specific, major task is being done.

Color coding is intended to illustrate the nature of the risk accrual for each major task within this phase of the process. Color code risk definitions are found in Chapter 2.

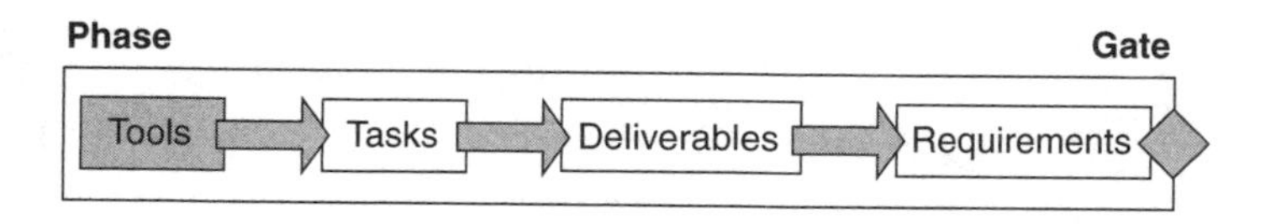

The tools, methods, and best practices that enable the tasks associated with fulfilling the Evaluate deliverables include the following:

- Financial modeling and forecasting tools such as NPV, ECV, ROI analysis, and Monte Carlo Simulation
- Business case development and valuation methods
- Portfolio balancing methods
- RWW Analysis
- Pugh Concept Evaluation & Selection Method

TABLE 4.9 Sample Evaluate Phase Tools Scorecard

1	2	3	4	5	6	7
SSFM Tool	**Quality of Tool Usage**	**Data Integrity**	**Results Versus Requirements**	**Average Tool Score**	**Data Summary, Including Type and Units**	**Task**
Product portfolio architecting methods						Generate candidate portfolio architectures
Pugh Concept Evaluation & Selection Process; portfolio balancing methods						Evaluate portfolio architecture candidates and select a best portfolio architecture; conduct the Pugh Process
Business case development and financial modeling methods (including Monte Carlo simulations)						Create preliminary business cases for the elements of each candidate portfolio architecture
RWW Analysis method						Conduct RWW Analysis on candidate portfolio architectures
Portfolio architecture FMEA						Conduct risk analysis on each candidate portfolio architecture (FMEA)
Project management methods						Develop Activate project plans and project risk analysis

- Project management tools including Monte Carlo Simulation for statistical cycle-time design and forecasting, and FMEA for cycle-time risk assessment.

The Scorecard, as shown in Table 4.9, is used by marketing teams who apply tools to complete tasks.

Columns 1 and 7 align the tool to a specific task and its requirement for justifying the use of the tool. (Never use a tool if you can't justify its ability to fulfill a task.)

Quality of Tool Usage, Data Integrity, and Results Versus Requirements are scored on a scale of 1 to 10 using the simple scoring templates provided in Chapter 2.

Average Tool Score illustrates overall tool quantification across the three scoring dimensions. (A high ATS indicates that a task is being underwritten by proper tool usage.)

Data Summary is intended to illustrate the nature of the data and the specific units of measure (attribute or continuous data, scale or specific units of measure).

Activate Phase Tools, Tasks, and Deliverables

As in the earlier phases, producing the final Activate phase deliverables requires an investment in numerous, detailed inbound marketing tasks enabled by specific tools, methods, and best practices. Their integration ensures that the right data is being developed and summarized to fulfill the major requirements at the Activate gate review.

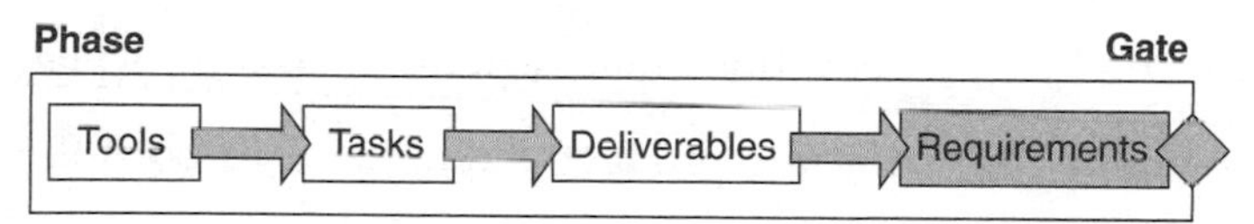

This phase has two major requirements guiding the Activate gate deliverables. The first requirement is to *rank the order of activation projects*. This involves balancing the value to deploy specific commercialization projects with specific timing intervals (when to wait versus when to take action). The second and last major requirement is the *timing for activation* of projects, based on a risk-balanced commercialization project control plan. This requirement encompasses the proper timing and use of the company resources and core competencies to safely develop and transfer technology and activate commercialization projects to sustain the growth goals.

Figure 4.17 summarizes the tools, tasks, and deliverables used during the Activate phase:

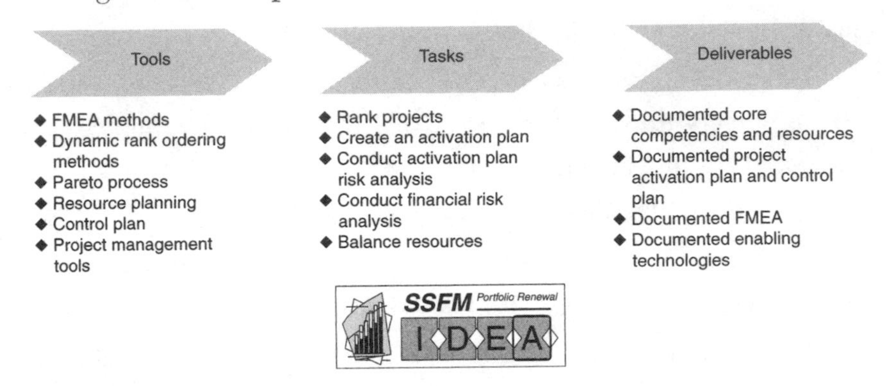

FIGURE 4.17 The tools-tasks-deliverables in the Activate phase of the IDEA process.

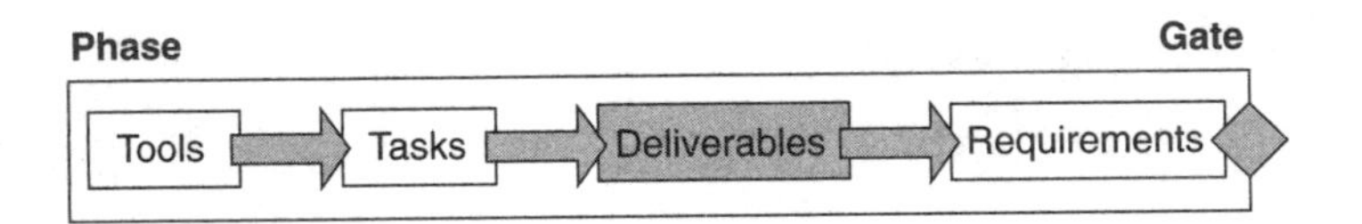

The four major deliverables required for the Activate gate are as follows:

- Documented availability, readiness, and deployment of core competencies and resources
- Documented project activation timing schedule and control plan
- Documented FMEA on the product and/or services portfolio activation control plan
- Documented enabling technology maturity and readiness

No supporting deliverables are required for the Activate phase.

The scorecard, as shown in Table 4.10, is used by marketing gatekeepers who manage risk and make functional gate decisions for a

TABLE 4.10 Sample Activate Gate Deliverables Scorecard

1	2	3	4	5	6
SSFM Gate Deliverable	**Grand Average Tool Score**	**% Task Completion**	**Tasks Results Versus Gate Deliverable Requirements**	**Risk Color (R, Y, G)**	**Gate Requirement(s) (General Phase Requirement)**
Documented dynamic rank order of projects; Pareto charts					Have the projects within the new portfolio architecture rank-ordered for risk-balanced activation (rank order of activation projects)
Documented project activation timing schedule and control plan; Monte Carlo Simulation documents					Define the timing for activation of projects, based on a risk-balanced commercialization project control plan (timing of activation)
Documented availability, readiness, and deployment plan of core competencies and resources to support the flow of activated projects					Have the proper balance of resources aligned with the projects slated for activation
Documented project activation FMEA					Have a risk assessment of the activation plan
Documented technology development and transfer control plan and summary data					Have a preliminary technology risk assessment for new platforms and modular technologies and their supply chains
Revised Activate project plans and risk analysis					Make the Activate project plans and risk analysis up-to-date

specific project as part of the portfolio of projects being conducted by the marketing organization.

Columns 1 and 6 align the gate deliverable to a gate requirement. Each deliverable is justified as it contributes to meeting a gate requirement. (Never produce a deliverable if you can't justify its ability to fulfill a gate requirement.)

Grand Average Tool Score (GATS) illustrates aggregated tool quantification across the three scoring dimensions. (A high GATS indicates that a group of tasks is underwriting a gate deliverable.)

% Task Completion is scored on a 10 to 100% scale. This measure is critical if you want to understand how completely a group of related tasks are fulfilling a major gate deliverable.

Color coding is intended to illustrate the nature of the risk accrual for each major deliverable within this phase of the process. Color code risk definitions are found in Chapter 2.

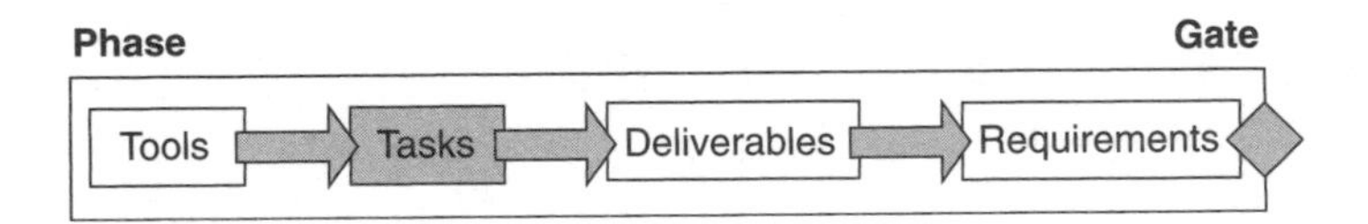

The five tasks to be performed within the Activate phase that will produce the deliverables are as follows:

1. Rank projects for activation priority and strategic value.
2. Create a project activation timing plan.
3. Conduct a risk analysis on the activation plan.
4. Conduct a risk analysis on the portfolio for financial performance against growth goals in light of the activation timing and plan.
5. Balance resources across the project activation plan.

The scorecard, as shown in Table 4.11, is used by marketing project team leaders who manage major tasks and their time lines as part of their project management responsibilities.

TABLE 4.11 Sample Activate Phase Task Scorecard

1	2	3	4	5	6
SSFM Task	**Average Tool Score**	**% Task Completion**	**Task Results Versus Gate Requirements**	**Risk Color (R, Y, G)**	**Gate Deliverable(s)**
Rank projects for activation priority and strategic growth value contribution over time					Documented dynamic rank order of projects; Pareto charts
Create project activation timing plan; conduct Monte Carlo simulations					Documented project activation timing schedule and control plan; Monte Carlo Simulation documents
Balance resources across the project activation plan					Documented availability, readiness, and deployment plan of core competencies and resources to support the flow of activated projects
Conduct risk analysis on portfolio performance against growth goals in light of activation timing alternatives					Documented project activation FMEA
Develop technology development and transfer control plans and supporting data sets					Documented technology development and transfer control plan and summary data
Update Activate project plan and risk analysis					Revised Activate project plans and risk analysis

Columns 1 and 6 align the task to a specific gate deliverable for justifying the task. (Never conduct a task if you can't justify its ability to produce a gate deliverable and fulfill a gate requirement.)

Average Tool Score (ATS) illustrates overall tool quantification across the three scoring dimensions. (A high ATS indicates that a task is being underwritten by proper tool usage.)

% Task Completion is scored on a 10 to 100% scale. This measure is critical if you want to understand how completely each specific, major task is being done.

Color coding is intended to illustrate the nature of the risk accrual for each major task within this phase of the process. Color code risk definitions are found in Chapter 2.

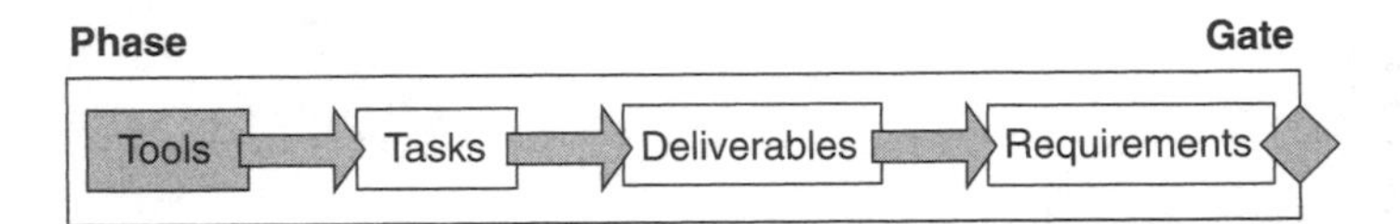

The tools, methods, and best practices that enable the tasks associated with fulfilling the Activate deliverables are as follows:

- FMEA methods
- Dynamic rank ordering methods
- Pareto process
- Resource planning and budgeting methods
- Control plan design methods
- Project planning and management tools such as Monte Carlo Simulation for statistical cycle-time design and forecasting, and FMEA for cycle-time risk assessment

The scorecard, as shown in Table 4.12, is used by marketing teams who apply tools to complete tasks.

Columns 1 and 7 align the tool to a specific task and its requirement for justifying the use of the tool. (Never use a tool if you can't justify its ability to fulfill a task.)

TABLE 4.12 Sample Activate Phase Tools Scorecard

1	2	3	4	5	6	7
SSFM Tool	**Quality of Tool Usage**	**Data Integrity**	**Results Versus Requirements**	**Average Tool Score**	**Data Summary, Including Type and Units**	**Task**
Dynamic rank ordering, methods Pareto ranking process						Rank projects for activation priority and strategic growth value contribution over time
Project management, cycle-time modeling, and control planning methods						Create project activation timing plan; conduct Monte Carlo simulations
Resource planning and budgeting methods						Balance resources across the project activation plan
Portfolio financial FMEA						Conduct risk analysis on the portfolio performance against growth goals in light of activation timing alternatives
Technology development and transfer control planning methods						Develop technology development and transfer control plans and supporting data sets
Project management methods						Update Activate project plan and risk analysis

Quality of Tool Usage, Data Integrity, and Results Versus Requirements are scored on a scale from 1 to 10 using the simple scoring templates provided in Chapter 2.

Average Tool Score (ATS) illustrates overall tool quantification across the three scoring dimensions. (A high ATS indicates that a task is being underwritten by proper tool usage.)

Data Summary is intended to illustrate the nature of the data and the specific units of measure (attribute or continuous data, scale or specific units of measure).

Summary

You now have a good foundation for how to apply Six Sigma for inbound marketing to drive the strategic process of portfolio renewal. The Identify-Define-Evaluate-Activate structure provides a strong linkage to the technical departments in your firm. Both disciplines now have a common language to communicate and translate customer requirements into business and product (or services) requirements. Plus, outputs from this process become inputs for the market opportunity assessment required to exit the first phase gate of the commercialization process.

Recall that it is essential for your product and services offering portfolio to be evaluated and refreshed. To unleash the power of Six Sigma—to manage by fact—you need to ensure that your portfolio renewal activities adhere to a process, a set road map, so that the team understands its requirements and the tools, tasks, and deliverables needed to successfully fulfill those requirements. IDEA gives you that structure to ensure measurable and predictable results that fulfill the gate requirements. You may choose to call your process phases by different names—that's fine. *What you do and what you measure* are what really matter.

The major steps of the portfolio renewal process are as follows:

- **Portfolio renewal process phases and gates:** Tools, tasks, deliverables, requirements, project management, and performance scorecards for the IDEA process

- **Market and segment identification:** Where are the boundaries where profit potential reside for application of your core competencies?
- **Opportunity scoping and mapping:** SWOT, aligning core competencies to market potential, behavior-based process map (internal perspectives, empathy and experiences, usage trends and habits, process behaviors)
- **"Over-the-horizon" VOC gathering for portfolio typing and characterization:** Interviewing, KJ, NUD screening (market needs within and across segments)
- **Statistical customer survey design and analysis:** Survey design, t-Tests and customer-ranked NUD requirements (finding statistically significant segments)
- **Portfolio typing and characterization process:** Aligning the right type of offering portfolio with each significant segment
- **Product portfolio alternative architecting:** Portfolio concept generation for products and services
- **Technical platform NUD requirements/technology HOQ/technology platform architecting**
- **Portfolio valuation and risk analysis:** RWW, FMEA, aggregate NPV, ECV, dynamic rank ordering, other available tools
- **Portfolio evaluation and selection:** Pugh Process and other methods
- **Commercialization project activation and control plans:** Project management and family plan deployment scheduling, adaptive risk management, resource allocation and balancing

Now that you understand the strategic marketing process, it is time to switch gears to a tactical marketing process and examine how best to apply Six Sigma to the activation of a commercialization project.

5

SIX SIGMA IN THE TACTICAL MARKETING PROCESS

Inbound Marketing for Commercialization of Products and Services

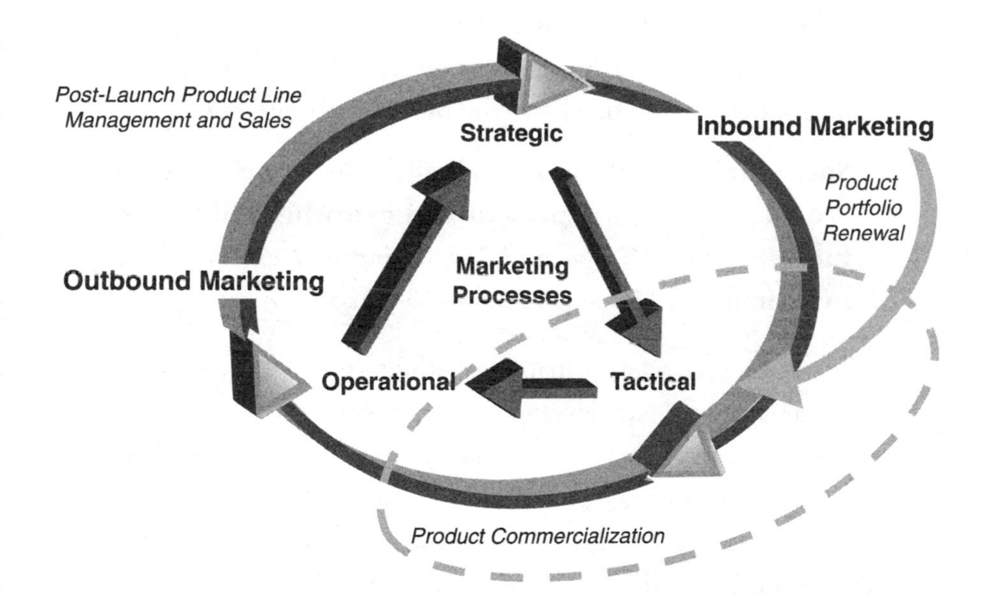

Tactical Marketing Process: Commercialization

The portfolio renewal process (discussed in Chapter 4, "Six Sigma in the Strategic Marketing Process") has initiated a commercialization project of a product and/or service to get ready for the marketplace. The commercialization approach we propose provides a road map to integrate customer requirements and translate their input into internal, actionable language. This translation becomes the common language between marketing and the other supporting disciplines (including primarily the technical community, but also finance and field operations). The process gathers Voice of the Customer (VOC) input (using the customer's words) and converts it into an internally actionable language, regardless of discipline, without losing the original meaning. This process focuses on applying Six Sigma concepts to marketing's role throughout the commercialization process of readying an offering for the market itself. This process assumes that the many keys supporting marketing deliverables (such as marketing collaterals, advertising campaign, public relations plan, and training) from your current processes are integrated into this process. The objective is to suggest a Six Sigma foundation from which to improve marketing's ability to meet its commercialization goals and to customize its approach to meet specific business requirements.

Successful organizations manage the future; others are managed by the present and overwhelmed by the future. (From *Design and Marketing of New Products* by Urban and Hauser, ISBN 0-13-201567-6)

Markets change faster than marketing. The importance of speed and accuracy in product development escalates with increasing global competition. Increasing rates of change in technology, buyer preferences, channel leverage, and the regulatory environment dictate shorter product life cycles and increase financial pressure on sales growth, profits, and shareholder value. You have two choices—innovate or harvest. Innovation is profitable, but it carries significant risk of failure. This chapter outlines a commercialization process for *innovation with lower risk* of failure for both products and services. Proven tools, process methodologies, and program management integrate learnings from best practices of organizations that enjoy consistent growth in sales, profits, and shareholder value.

Commercialization is an exciting time for a business—representing its growth cycle. The portfolio renewal team has completed the latest cycle of their strategic work of architecting new elements to add to the product and/or services portfolio. They have ranked and prioritized a balanced and linked set of new ideas and market opportunities that have passed the IDEA Gate Reviews. The ideas and opportunities have been merged into candidate commercialization projects. Now it is time to activate a serial or parallel flow of new commercialization projects from the renewed offering portfolio architecture. Small companies, or a division, may activate one product or product family. Larger companies may activate a staggered series or parallel group of product commercialization projects. This "loading" of the offerings pipeline means it is time to convert preliminary ideas into numerous alternative product and/or services concepts as opportunities convert into precise requirements for each offering being commercialized. The goal is to launch the right product and/or service based on a credible set of requirements.

This tactical phase is initiated, and a commercialization team is formed. Ideally, a multifunctional team is constructed around a core of mainly marketing, design, and production professionals who will do the tactical work of commercializing the product, readying it for launch. As the development process approaches launch, additional disciplines, particularly representative operational professionals such as customer service and training, may join the commercialization team to smooth the transition from a tactical commercialization process to an operational process.

Commercialization of a professional services offering may require fewer disciplines to be represented on the multifunctional team, while the commercialization of a "solution" (combining technology and services) may require a broader array of disciples. Regardless of the offering being readied for launch, the Six Sigma approach is the same and can be applied to either a tangible product or a professional services offering. Throughout this book, the word "product" refers to a company's generic "offering" to the marketplace and represents both a tangible product and a services offering.

Commercialization Phases and Gates: Introduction to UAPL

The use of phases and gates is a time-tested approach to control the timing and quality of marketing deliverables, tasks, and tools, and to manage risk throughout the commercialization process. Marketing work is broken into "designed" clusters that can be planned and managed to produce predictable quality results—on time and within budget. Allowance must be made for surprises (unknowable events) that can disrupt the planned work. Adaptive deployment of planned marketing tasks must be under control if you are to be a good steward of the corporate investment being made in the commercialization project. Make every reasonable effort to design a plan of marketing work and then live by that plan. You can then make the necessary adjustments to that plan as circumstances force you to do so. However, making frequent changes in a planless commercialization process eventually leads to chaos. Following a set of steps may slow things down at first, but you're slowing down to speed up. This comes with experience. Tasks, methods, and tools can also prevent common human behaviors from undermining the process. For example, linear thinking may rush a team to select a suboptimal product concept without adequate deliberation.

This chapter examines the tactical commercialization process through the lens of a marketing professional focused simply on the support that Six Sigma can provide. Hence, the tools, tasks, and deliverables discussed in this chapter center on the Six Sigma components that a marketer employs in this tactical process. Marketing professionals perform many activities and produce (or manage the development of) multiple deliverables (such as marketing collaterals, advertising campaign, public relations plan, training) that are not reviewed in this book. The objective is to provide a Six Sigma foundation from which to improve marketing's ability to meet its commercialization goals and to customize its approach to meet specific business requirements. Figure 5.1 shows the essential macro tasks for marketing during commercialization.

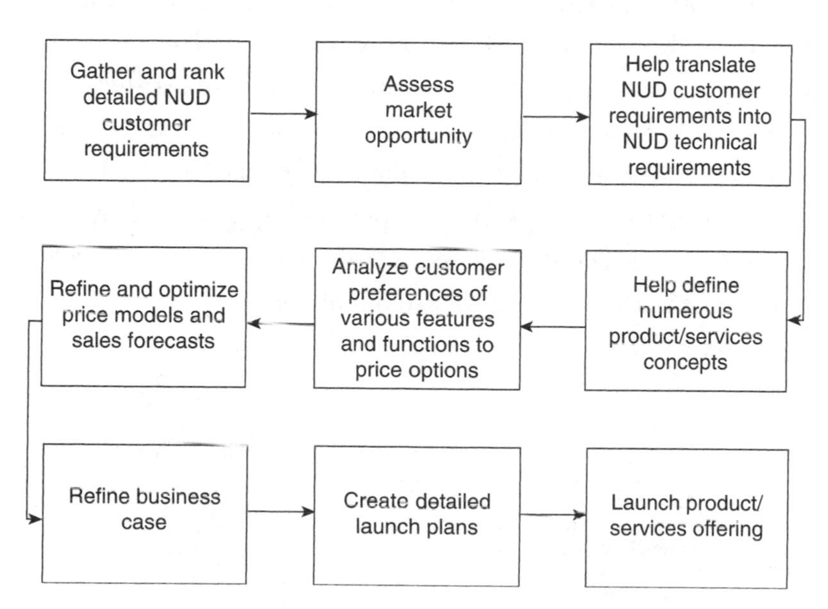

FIGURE 5.1 Macro tasks for commercialization.

It is interesting to compare these macro steps to those from the strategic portfolio renewal process in the preceding chapter. Note how marketing tasks shift from general to very specific as a product and/or service is commercialized. Regardless of who in the firm performs these activities, the marketing tasks become more tactical as the offering approaches launch.

If the team lacks the required strategic portfolio work to initiate its commercialization project, however, the team must step back and gather the prerequisite information. For example, you should develop a market opportunity assessment that defines targeted segments to assess attractiveness. Then compare that to an internal assessment to determine strategic fit and organizational capabilities. A financial risk-versus-return analysis gives the gatekeepers the critical information to help make investment decisions. The goal is to enter a commercialization project with enough market research to inform a "go/no go" decision before more investments are made in activities such as gathering VOC data.

In Chapter 4, Figure 4.3 illustrates the integration of marketing and technical functions in the context of inbound and outbound marketing arenas. Although marketing typically drives these three processes (strategic, tactical, and operational—see Figure 5.2), it needs to ensure that other supporting functional groups (such as its technical counterparts) provide respective input and deliverables throughout. The tactical process environment for marketing takes input from the strategic processes to initiate the preparation (or commercialization) process of a product or service for launch.

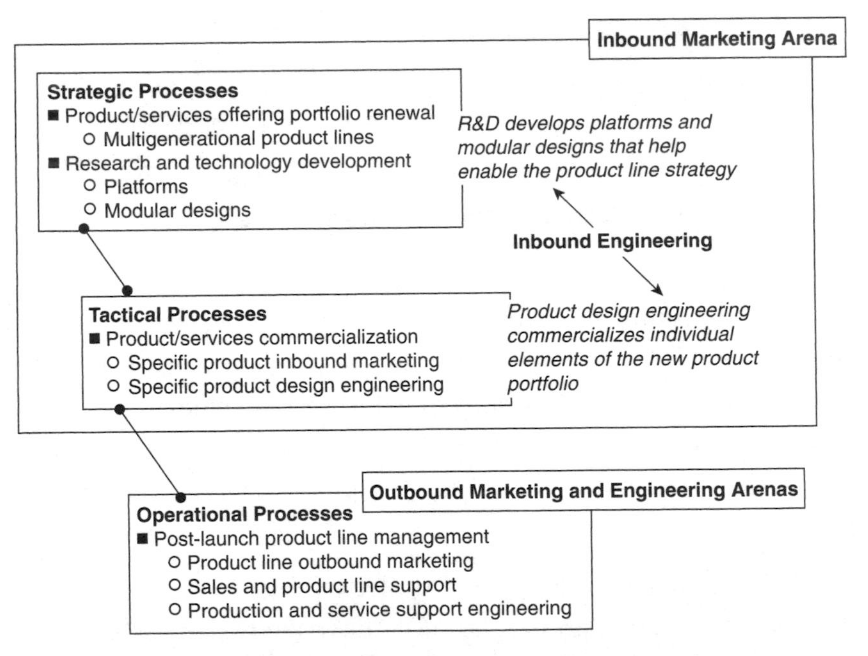

FIGURE 5.2 Integrated marketing and technical functions for tactical.

In the tactical marketing process environment, we have identified a generic commercialization process that is segmented into four distinct phases. Your company will almost certainly have its own names for the phases of commercialization. If so, the deliverables and tasks from this book's approach can be redistributed among the named phases your company uses. The four-phase commercialization approach we propose is Understand-Analyze-Plan-Launch. This generic UAPL marketing process for specific commercialization projects is defined as follows:

1. **Understand** specific customer need details that must be precisely translated into a set of clear technical product requirements that sustain the value of the preliminary business case.
2. **Analyze** customer preferences for and competitive threats to the features and functions of the evolving product concept. Analysis is used to refine the value proposition for price and sales forecasting and offering positioning within the context of the brand.
3. **Plan** the linkage between marketing and sales process details, with other supporting functional input, to successfully communicate value. Position and launch the new offering in fulfillment of the maturing business case.
4. **Launch** the new offering under a rigorously defined launch control plan with appropriate measures of marketing and sales performance that help prevent failures and mistakes (prevention of escaping marketing "defects").

FIGURE 5.3 The Understand gate review.

UAPL is a generic phase-gate vocabulary (see Figure 5.3). It names the commercialization phases according to what work generally is executed within each of them. Elements of the market opportunity assessment can be further developed into the business case.

The following sections align the tools within each phase of the unique UAPL inbound marketing method to commercialize an offering. The suggested tools draw from the complete set of Six Sigma/Lean tools readily available today; no "new" tools are introduced. Because this book is an executive overview, individual tool descriptions and guidelines on how to use them fall outside the scope of this book. However, each Deliverables, Tasks, and Tools section contains a sample scorecard to show the hierarchical linkage of the tools-tasks-deliverables-requirements combination. In addition, it is important to mention that commercialization projects, like any Six Sigma project, may or *may not* use all the tools aligned within a given phase; it depends on the project's context and complexity. Hence, it is important to understand how and when to use the tools to ensure that *the right tool is used at the right time to answer the right question.*

The Understand Phase

The objective of the Understand phase is for marketing to gain a precise understanding of customer needs and preferences. The marketing team drives alignment between market/segment opportunities and the product IDEA. They apply a specific set of marketing tasks and tools to build a precise and stable array of requirements, based on the fresh gathering of VOC data for the specific product (or services offering) being commercialized.

Marketing collaborates with the technical professionals on the team to translate *customer needs* (precisely defined, detailed, New, Unique, and Difficult [NUD] needs you can fulfill) into clear, precise *requirements* (a precise definition of NUD technical requirements with targets and ranges of acceptance) to avoid developing the wrong product. Marketing helps develop numerous offering concept alternatives for assessment once the requirements are clearly defined.

Recall that the Six Sigma structure uses a phase-gate approach, wherein a phase is a period of time designated to conduct work to produce specific results that meet the requirements for a given project.

This book takes the view that every cycle of the tactical commercialization process is a project with distinct phases and gates. A gate is a stopping point where you review results against requirements for a bounded set of commercialization tasks. The requirements govern the criteria with which deliverables are judged successful or unsuccessful. A phase is normally designed to limit the amount of risk that can accrue before a gatekeeping team assesses the summary data that characterizes the risk of going forward. A system of phases and gates is how the product commercialization process is put under control by defining the control plan for the inbound marketing processes.

Requirements for the Understand Gate

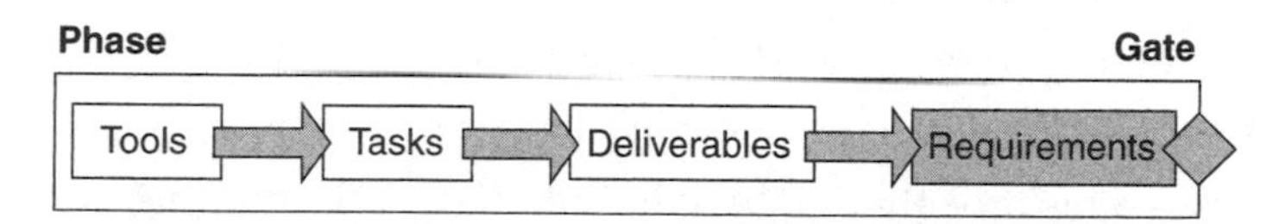

The first phase of UAPL requirements builds on the work completed in the strategic portfolio definition and development process to activate a commercialization project. Hence, the requirements are five-fold:

- Documented customer requirements
- Documented product requirements
- Documented superior product concept(s)
- Refined and updated Real-Win-Worth (RWW) Analysis
- Refined and updated business case

In preparation for a phase-gate review, marketing ensures that the product requirements incorporate input from its technical counterparts. Marketing needs to jointly develop the last three phase-gate requirements (product concept, RWW, and business case) with the technical team and other supporting functional groups, such as finance. Elements of the market opportunity assessment can be further developed into the business case.

Understand Phase Tools, Tasks, and Deliverables

Producing the Understand phase deliverables requires an investment in numerous, detailed inbound marketing tasks enabled by specific tools, methods, and best practices. Their integration ensures that the right data is being developed and summarized to fulfill the major requirements at the Understand gate review.

The Understand phase is critical to anchor the entire UAPL process. Gathering all the facts and taking the time to plan and accurately translate requirements are vital to establishing the correct course for the entire commercialization process. Hence, this phase has many activities. Skimping on these tasks or deliverables could lead to a misaligned offering or launch. The tools help the tasks produce the right deliverables to meet the phase-gate requirements and ensure measurable results. The following list summarizes the tools, tasks, and deliverables for the Understand phase (see Figure 5.4).

Tools

- Product category data mining
- VOC gathering methods
- Focused VOC data processing
- Product category SWOT
- Market perceived quality profile
- Porter's 5 Forces Analysis
- Customer behavioral dynamics map
- Competitive benchmarking
- Value chain analysis
- Customer value map
- Key events timeline
- Competitive position analysis
- Won-Lost Analysis
- QFD and HOQ methods
- Concept generation tools
- Pugh concept
- RWW Analysis
- Business case modeling
- Project management tools

Tasks

- Define product fit
- Define fit with marketing innovation
- Define customer types
- Create customer interview guide
- Gather and translate VOC
- Document NUD
- Construct behavioral dynamics map
- Benchmark competitors
- Help translate into technical requirements
- Help evaluate concepts
- Define qualitative and quantitative
- Update marked perceived quality profiles
- Analyze value chain
- Develop product category
- Analyze Won-Lost
- Analyze competitive position
- Update RWW Analysis
- Refine business case
- Create ANALYZE project plan

Deliverables

- Documented business case goals
- Documented core competencies
- Documented innovation strategy
- Segmented markets
- Documented customer requirements
- Documented product requirements
- Documented superior product concept
- Created marketing critical parameters
- Updated RWW Analysis
- Updated business case
- Documented ANALYZE project plan

Supporting Deliverables

- NUD customer requirements
- Competitive benchmarking
- Product-level House of Quality
- Updated perceived quality profile
- Updated Porter's 5 Forces chart
- Updated SWOT matrix

FIGURE 5.4 The tools-tasks-deliverables for the Understand phase.

Understand Gate Deliverables

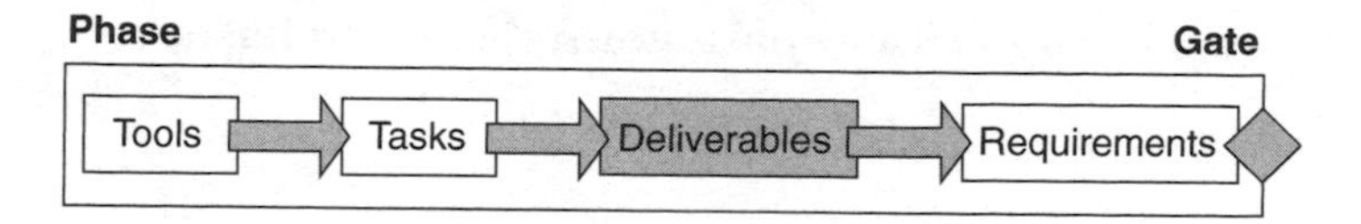

The deliverables for the Understand gate fall into two categories—major and supporting deliverables. Both are critical outputs to this phase and should be inspected for completeness and accuracy. The many major deliverables are all equally important; no one output should be overlooked or incomplete. The 11 Understand major deliverables are as follows:

- **Documented business case goals,** including financial targets and tolerances, and what percentage the new offering would contribute to the overall portfolio financial requirements for a specified period of time. Business case development may be supported by finance, but marketing often owns the content and is accountable for its assumptions.
- **Documented marketing core competencies**, their availability, and readiness at time of launch.
- **Documented marketing innovation strategy** for the project.
- **Identified market segments** that this product and/or service will serve, represented as micro profit pools.
- **Documented specific NUD customer requirements (VOC):** The *new* needs customers have not asked you or any of your competitors to fulfill in the past. The *unique* needs customers have expressed that your competitors are fulfilling, but you are not. The *difficult* requirements are explicit opportunities to address if you can use your core competencies to overcome associated hurdles.
- **Documented product requirements** that define value and customer need dynamics.

- **Documented superior product concept.**
- **Create marketing critical parameters and database.**
- **Updated RWW Analysis.**
- **Updated business case.**
- **Analyze phase project management and risk plan.** Although the project management activities may be supported by another discipline (such as the technical community), marketing often owns the overall management of the commercialization project and is accountable for (at a minimum) marketing's deliverables and overall risk mitigation planning for commercialization.

The Six Sigma term *critical parameters* refers to *fundamental* or absolutely necessary elements needed to manage the offering throughout its life cycle against its plan. Critical parameters are unique to your business and the offering. Critical Parameter Management (CPM) identifies those fundamental requirements and measures critical functional responses to design, optimize, and verify the capability of an offering and its supporting value chain processes to meet requirements. The requirements need to prioritize and integrate what is critical to both the internal and external demands (VOC and Voice of the Business [VOB]).

A system of critical parameters that control the dynamics (cause-and-effect relationships) of the execution of the marketing and sales plans, as well as the product launch, are structured and linked. Marketing, sales, and launch process noises (specific sources of disruptive variation) are identified. Plans are made robust to these sources of variation. A specific set of controllable marketing and sales parameters are identified to ensure that their cause-and-effect relationships are statistically significant and can move mean launch performance variables to hit their required targets during the ramp-up to steady-state sales and support rates. These requirements and measured results and their governing marketing and sales parameters are the elements that are used to structure launch, customer relationship, and channel management control plans.

The marketing and sales critical parameters reside within the following: channel management strategy and plan, distribution plan, sales support plan, training plan (both internal and external), sales promotion plan, and communications plan. The marketing communication plan should consider advertising, public relations, trade shows, and collaterals (customer brochures, direct mailers, or "leave behinds").

The six supporting Understand gate deliverables are as follows:

- **NUD customer requirements,** which take into account customer ID matrix, customer interview guide, and KJ diagrams (of both images and requirements).
- **Competitive benchmarking data and trend analysis.**
- **Product level HOQ** for the NUD customer requirements and customer ranks on the NUDs, as well as the customer benchmarking ranks on NUD customer requirements.
- **Updated market perceived quality profile** and gap analysis for the specific product category for this project.
- **Updated Porter's 5 Forces Analysis** and specific product risk profile.
- **Updated SWOT Analysis** and specific product risk profile.

The sample scorecard, as shown in Table 5.1, is used by marketing gatekeepers who manage risk and make functional gate decisions for a specific project as part of the portfolio of projects being conducted by the marketing organization.

Columns 1 and 6 align the gate deliverable to a gate requirement. Each deliverable is justified as it contributes to meeting a gate requirement. (Never produce a deliverable if you can't justify its ability to fulfill a gate requirement.)

TABLE 5.1 Sample Understand Gate Deliverables Scorecard

1	2	3	4	5	6
SSFM Gate Deliverable	**Grand Average Tool Score**	**% Task Completion**	**Tasks Results Versus Gate Deliverable Requirements**	**Risk Color (R, Y, G)**	**Gate Requirements**
Document illustrating alignment of target with segmentation scheme					Define select targets (focused opportunities) using attraction criteria (such as revenues and growth rate)
Updated marketing core competencies and capabilities assessment					Assess existing (or missing) organizational capabilities to exploit targeted segments profitably
Update competitive analysis					Characterize competitive trends within six months and describe how advantage will be achieved
Update market perceived quality profile					Develop detailed analysis of market perceived quality profile and analyze gaps validated by VOC
Documented update of Porter's 5 Forces Analysis					Update opportunity/risk assessment
Documented update of SWOT Analysis					

Documented customer requirements					Have ranked and structured sets of NUD customer requirements by segment and target. Must present customer requirements ranking survey results.
Documented product requirements					Document product level NUD and ECO requirements
Documented product level HOQ					Show how customer requirements are translated into ranked NUD technical requirements
Documented product concepts					Present alternative product concepts that also meet customer requirements
Documented superior product concept					Define superior product concept
Documented marketing critical parameters					Show critical customer NUD needs that are clear, measurable, and value-adding differentiators to underwrite business case
Updated business case					Business case must be product concept-specific and in alignment with business objectives and goals
Updated RWW Analysis					Estimate concept financial potential against current risk assessment
Document Analyze phase project management and risk plan					Illustrate Analyze phase project management and risk plan

Grand Average Tool Score (GATS) illustrates aggregated tool quantification across the three scoring dimensions. (A high GATS indicates that a group of tasks is underwriting a gate deliverable.)

% Task Completion is scored on a 10 to 100% scale. This measure is critical if you want to understand how completely a group of related tasks are fulfilling a major gate deliverable.

Color coding is intended to illustrate the nature of the risk accrual for each major deliverable within this phase of the process. Color code risk definitions are found in Chapter 2, "Measuring Marketing Performance and Risk Accrual Using Scorecards."

Understand Phase Tasks

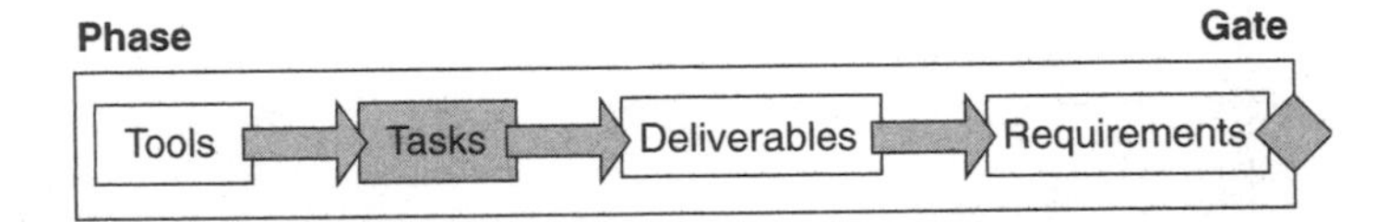

Given the many deliverables required to complete the first UAPL phase, the tasks within the Understand phase parallel and match the outputs. They include the following:

1. Conduct a Goals, Objectives, Strategies, Plans, and Actions (GOSPA) analysis, a planning methodology.
2. Define how this product currently fits within the business strategy and financial goals of the product portfolio (current priority check).
3. Define the fit with the marketing innovation strategy.
4. Define specific customer types and where they are located.
5. Create a customer interview guide.
6. Gather and translate specific VOC data.
7. Document NUD customer needs.
8. Construct specific customer behavioral dynamics maps.

9. Conduct competitive benchmarking against the NUD customer requirements.
10. Help translate customer requirements into technical requirements.
11. Help generate and evaluate concepts that are candidates to fulfill the NUD.
12. Define the qualitative and quantitative elements of value that the concept provides to substantiate the specific product opportunity.
13. Update market perceived quality profiles.
14. Conduct customer value chain analysis.
15. Develop product category-specific key event time line.
16. Conduct Won-Lost Analysis.
17. Generate competitive position analysis.
18. Update RWW Analysis.
19. Refine and update the business case.
20. Create an Analyze phase marketing project plan and risk analysis.

The sample scorecard, as shown in Table 5.2, is used by marketing project team leaders who manage major tasks and their time lines as part of their project management responsibilities.

Columns 1 and 6 align the task to a specific gate deliverable for justifying the task. (Never conduct a task if you can't justify its ability to produce a gate deliverable and fulfill a gate requirement.)

Average Tool Score (ATS) illustrates overall tool quantification across the three scoring dimensions. (A high ATS indicates that a task is being underwritten by proper tool usage.)

TABLE 5.2 Sample Understand Phase Task Scorecard

1	2	3	4	5	6
SSFM Task	**Average Tool Score**	**% Task Completion**	**Task Results Versus Gate Requirements**	**Risk Color (R, Y, G)**	**Gate Deliverable**
Conduct GOSPA analysis					Document illustrating alignment of target with segmentation scheme
Revise marketing core competencies and capabilities assessment					Updated core competencies and capabilities assessment
Revise Porter's 5 Forces Analysis					Documented update to Porter's 5 Forces Analysis
Revise SWOT Analysis					Documented update of SWOT Analysis
Define specific customer types and where they are located					Documented customer requirements
Create interview guide					
Gather VOC data					
Conduct KJ Analysis to structure and rank data					
Document NUDs					
Construct customer behavioral dynamics map					
Develop and implement customer requirements ranking survey					
Help translate customer requirements into technical requirements					Documented product level HOQ

Conduct competitive benchmarking against NUDs					Update competitive analysis
Help generate and evaluate candidate concepts					Documented product concepts
Define qualitative and quantitative elements of value					Documented superior product concept
Convert eMPQP to Market Perceived Quality Profiles (MPQP) with VOC data					Update market perceived quality profile
Develop category key events time line					
Develop Won-Lost Analysis					Update market perceived quality profile/update competitive analysis
Construct competitive position analysis					
Conduct customer value chain analysis					Documented marketing critical parameters
Define how product fits business strategy/financial goals					Updated business case
Define fit with marketing innovation strategy					
Estimate concept financial potential against current risk assessment					Updated RWW Analysis
Create Analyze phase project plan and risk analysis					Document Analyze phase project management and risk plan

% Task Completion is scored on a 10 to 100% scale. This measure is critical if you want to understand how completely each specific, major task is being done.

Color coding is intended to illustrate the nature of the risk accrual for each major task within this phase of the process. Color code risk definitions are found in Chapter 2.

Tools That Enable the Understand Tasks

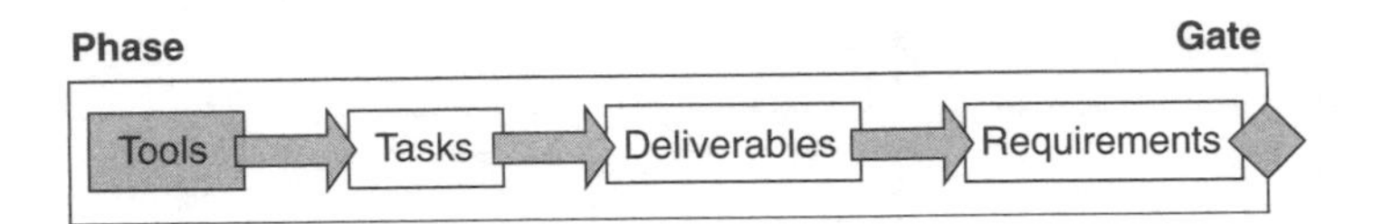

The tool set that supports the Understand tasks to provide the Understand deliverables includes the following software, methods, and best practices:

- **GOSPA** method to define your Goals, Objectives, Strategies, Plans, and Actions.
- **Product category data mining and analysis**, composed of secondary market research and data-gathering tools, economic and market (segment) trend forecasting tools, and statistical (descriptive, graphical, inferential, and multivariate) data analysis.
- **VOC gathering methods**, including primary market research and customer data gathering tools.
- **Focused VOC data processing best practices,** such as the KJ diagramming method and questionnaire and survey design methods.
- **Product category-specific SWOT** Analysis method.
- **Market perceived quality profile** and gap analysis method.

- **Porter's 5 Forces Analysis** method.
- **Customer behavioral dynamics** mapping methods.
- **Competitive benchmarking** tools.
- **Value chain analysis.**
- **Customer value map.**
- **Competitive position analysis.**
- **Won-Lost Analysis.**
- **QFD** and **HOQ** methods.
- **Concept generation** tools.
- **Pugh Concept** Evaluation & Selection Method.
- **RWW Analysis.**
- **Business case** development with financial pro forma models (such as Monte Carlo Simulation).
- **Project management tools,** including critical path time line, Monte Carlo Simulation for statistical cycle-time design and forecasting, and Failure Modes & Effects Analysis (FMEA) for cycle-time risk assessment.

The sample scorecard, as shown in Table 5.3, is used by marketing teams who apply tools to complete tasks.

Columns 1 and 7 align the tool to a specific task and its requirement for justifying the use of the tool. (Never use a tool if you can't justify its ability to fulfill a task.)

Quality of Tool Usage, Data Integrity, and Results Versus Requirements are scored on a scale of 1 to 10 using the simple scoring templates provided in Chapter 2.

Average Tool Score illustrates overall tool quantification across the three scoring dimensions. (A high ATS indicates that a task is being underwritten by proper tool usage.)

TABLE 5.3 Sample Understand Phase Tools Scorecard

1	2	3	4	5	6	7
SSFM Tool	**Quality of Tool Usage**	**Data Integrity**	**Results Versus Requirements**	**Average Tool Score**	**Data Summary, Including Type and Units**	**Task**
GOSPA						Conduct GOSPA analysis
Marketing skills and resources audit						Revise marketing core competencies and capabilities assessment
Porter's 5 Forces Analysis						Revise Porter's 5 Forces Analysis
SWOT						Revise SWOT Analysis
Customer location matrix						Define specific customer types and where they are located
Interview guide creation methods						Create interview guide
Customer selection matrix						Gather VOC data
KJ Analysis						Conduct KJ Analysis to structure and rank data
NUD screen						Document NUDs
Process mapping						Construct customer behavioral dynamics map
Survey development methods						Develop and implement customer requirements ranking survey

Quality function deployment						Translate customer requirements into technical requirements
Kano analysis						Conduct competitive benchmarking against NUDS
Brainstorming, TRIZ, mind mapping						Help generate and evaluate candidate concepts
Kano analysis, MPQ gap analysis, value chain analysis						Define qualitative and quantitative elements of value
MPQ comparison, MP price profile, value maps						Convert eMPQP to MPQP with VOC data
Key events time line						Develop category key events timeline
Won-Lost Analysis						Develop Won-Lost Analysis
Competitive position analysis template						Construct competitive position analysis
Value chain analysis						Conduct customer value chain analysis
GOSPA method/NPV analysis/pro forma cash flow analysis						Define how product fits business strategy/financial goals, and fit to marketing innovation strategy
GOSPA method						
NPV, EVA, IRR, B/E analysis, SWOT, Porter's 5 Forces						Estimate concept financial potential against current risk assessment
Project management methods/ RACI Matrix						Create Analyze phase project plan and risk analysis

Data Summary is intended to illustrate the nature of the data and the specific units of measure (attribute or continuous data, scale or specific units of measure).

As part of the overall project management of the commercialization process, the control, management of risk, and key decision-making are greatly enhanced when each commercialization phase clearly defines the requirements, deliverables, tasks, and enabling tools. Recall that as in the portfolio renewal process, any vague tools-tasks-deliverables-requirements combination erodes the ability to sustain growth. When these four elements disintegrate, so does your performance, eventually resulting in erratic growth from sources (see Figure 5.5).

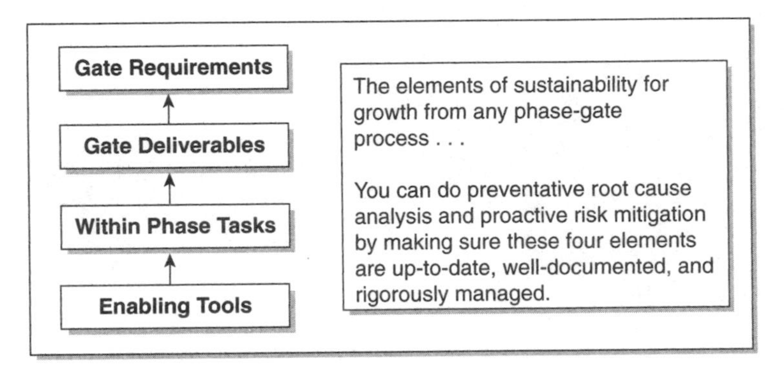

FIGURE 5.5 The Analyze gate review.

Using this approach, a distinct risk control plan for conducting tactical marketing functions in the product commercialization process can be structured.

The Analyze Phase

In this phase, marketing focuses on conducting a detailed analysis of market and segment trends, relevant competitive data, and specific customer preference data related to the product concept(s). The main objective is to refine the relative positioning of the offering concept—its price and sales forecast models that were originally estimated in the strategic portfolio renewal process.

To ensure a thorough analysis, marketing uses the data gathered from different sources in the Understand phase to analyze it; draw conclusions; and make positioning, price, and forecast recommendations to achieve the Analyze phase-gate requirements. Data on product concept variants is analyzed to assess key value and price drivers to refine price and sales forecast models. Customer feedback data on product/service concepts is evaluated to ensure that the offering's value is in balance with the branding strategy for the business. Marketing scrutinizes the competitive data and market trends to further refine the value proposition that will later be communicated in the marketing, advertising, promotional, and sales materials. Marketing also needs to consider the alignment of the maturing value proposition (form, fit, function, and features of the concept for a proposed price) to the entire brand. Based on the collective data analyzed, marketing works with the supporting disciplines to refine the business case.

Requirements for the Analyze Gate

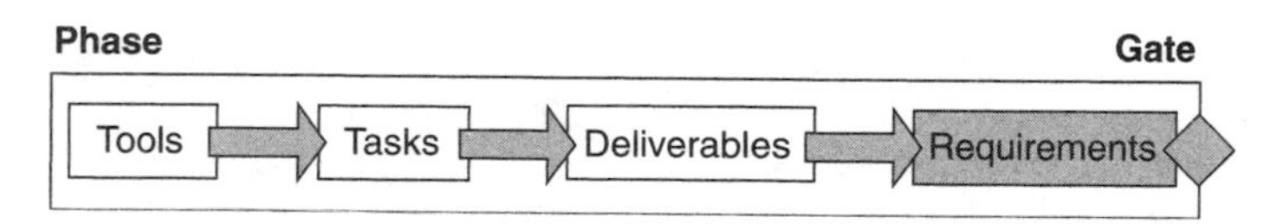

The second phase of UAPL requirements builds on the work completed in the Understand phase, which has eight requirements:

1. Finalized product requirements from customer testing.
2. Refined price model with Monte Carlo simulations.
3. Refined sales forecasts with Monte Carlo simulations.
4. Documented value proposition linked to brand.
5. Updated marketing and sales critical parameters and respective database(s).
6. Refined and updated RWW Analysis.
7. Refined and updated business case.
8. Plan phase project marketing plan, including a preliminary risk mitigation plan.

Analyze Phase Tools, Tasks, and Deliverables

Producing the Analyze phase deliverables requires an investment in numerous, detailed inbound marketing tasks enabled by specific tools, methods, and best practices. Their integration ensures that the right data is being developed and summarized to fulfill the major requirements at the Analyze gate review.

The Analyze phase provides the insight, from studying the facts, that guides the entire UAPL process. Analyzing all the facts before jumping to conclusions is important to establish a solid foundation for the entire commercialization process. Hence, this phase has a large number of activities. Skimping on these tasks or deliverables could lead to a misaligned offering or launch. The tools allow the tasks to produce the deliverables that meet the phase-gate requirements and ensure measurable results. The following list summarizes the tools, tasks, and deliverables for the Analyze phase (see Figure 5.6).

Tools

- Critical Parameter Management
- Customer based concept testing
- Conjoint studies and analysis
- Design of experiments
- Descriptive and inferential statistical data analysis
- ANOVA data analysis
- Regression and empirical modeling methods
- Multivariables studies
- Statistical Process Control
- Price modeling: market perceived price profile
- Sales forecasting
- Monte Carlo Simulation
- Sales channel analysis
- Brand development and management
- RWW Analysis
- Business case analysis
- Project management tools

Tasks

- Assess strategic fit
- Examine product development process capability
- Evaluate marketing capabilities
- Analyze sales force capabilities
- Conduct channel analysis
- Select pricing strategy/develop assumptions
- Create initial sales forecast
- Develop marketing plan
- Document critical risks, problems, and assumptions
- Update business case and financials
- Develop organizational communication plan
- Create PLAN project plan

Deliverables

- Critical Parameter Management database for marketing
- Organizational capabilities assessment
- Conjoint analysis results
- Customer preferences for form, fit, functions, and features at price points
- Price model (Monte Carlo simulations)
- Sales forecasts (Monte Carlo simulations)
- Marketing and sales channel strategy
- Brand positioning strategy for the value proposition
- Refined and updated RWW Analysis
- Refined and updated business case
- Documented PLAN phase project plans

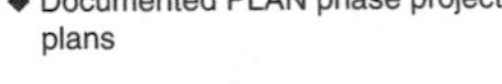

FIGURE 5.6 The tools-tasks-deliverables for the Analyze phase.

Analyze Gate Deliverables

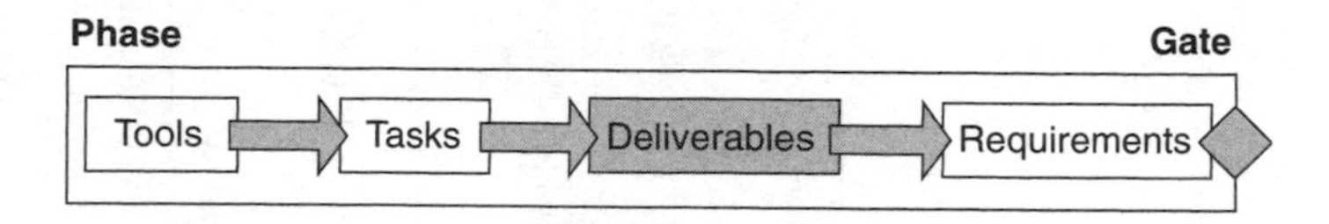

The deliverables at the Analyze gate are all major deliverables. They are critical outputs to this phase and should be inspected for completeness and accuracy. The many major deliverables are all equally important; no one output should be overlooked or incomplete. The 11 Analyze major deliverables are as follows:

- **Critical Parameter Management** database for marketing.
- **Organizational capabilities assessment.**
- **Conjoint analysis** results.
- **Customer preferences** for form, fit, functions, and features at price points.
- **Price model** (Monte Carlo Simulations).
- **Sales forecasts** (Monte Carlo Simulations).
- **Marketing and sales channel strategy.**
- **Brand positioning strategy** for the value proposition.
- **RWW Analysis** refined and updated.
- **Business case** refined and updated.
- **Plan phase project plan** developed and **risk plan** updated (Monte Carlo Simulation and FMEA on critical path).

The sample scorecard, as shown in Table 5.4, is used by marketing gatekeepers who manage risk and make functional gate decisions for a specific project as part of the portfolio of projects being conducted by the marketing organization.

TABLE 5.4 Sample Analyze Gate Deliverables Scorecard

1	2	3	4	5	6
SSFM Gate Deliverable	**Grand Average Tool Score**	**% Task Completion**	**Tasks Results Versus Gate Deliverable Requirements**	**Risk Color (R, Y, G)**	**Gate Requirement**
Updated marketing critical parameter management database					Have finalized product requirements from customer testing; update marketing and sales critical parameters and respective database(s)
Conjoint analysis and Monte Carlo Simulation results; customer documented preferences for form, fit, functions, and features at price points; documented price model					Refine and document updated price model with Monte Carlo simulations
Documented sales forecasts (Monte Carlo simulations)					Refine sales forecasts with Monte Carlo simulations
Documented brand positioning strategy for the value proposition					Have documented value proposition linked to the brand
Documented marketing and sales channel strategy					Document a marketing and sales channel strategy
Documented updates to the RWW Analysis					Refine and update RWW Analysis
Documented alignment of marketing objective with organizational goals					Show marketing objectives aligned with organizational goals
Documented updates to the business case					Refine and update business case
Documented Plan phase organizational capabilities assessment, project management, and risk mitigation plan					Document the Plan phase project plan, including a risk mitigation plan

Columns 1 and 6 align the gate deliverable to a gate requirement. Each deliverable is justified as it contributes to meeting a gate requirement. (Never produce a deliverable if you can't justify its ability to fulfill a gate requirement.)

Grand Average Tool Score (GATS) illustrates aggregated tool quantification across the three scoring dimensions. (A high GATS indicates that a group of tasks is underwriting a gate deliverable.)

% Task Completion is scored on a 10 to 100% scale. This measure is critical if you want to understand how completely a group of related tasks are fulfilling a major gate deliverable.

Color coding is intended to illustrate the nature of the risk accrual for each major deliverable within this phase of the process. Color code risk definitions are found in Chapter 2.

Analyze Phase Tasks

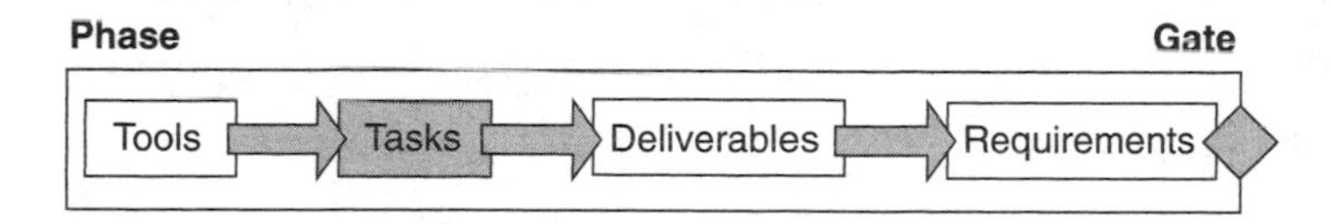

Given the many deliverables required to complete the second UAPL phase, the tasks within the Analyze phase parallel and match the outputs. They include the following:

1. Assess the strategic fit of the evolving product or service concept.
2. Examine the product development process capability to meet the launch date.
3. Evaluate marketing capabilities.
4. Analyze sales force capabilities.
5. Conduct channel analysis.
6. Select a pricing strategy and develop assumptions.
7. Create an initial sales forecast.
8. Develop a marketing plan.

9. Document critical risks, problems, and assumptions.
10. Update the business case and financials.
11. Develop an organizational communication plan.
12. Create a Plan project plan and update the risk plan.

The sample scorecard, as shown in Table 5.5, is used by marketing project team leaders who manage major tasks and their time lines as part of their project management responsibilities.

Columns 1 and 6 align the task to a specific gate deliverable for justifying the task. (Never conduct a task if you can't justify its ability to produce a gate deliverable and fulfill a gate requirement.)

Average Tool Score (ATS) illustrates overall tool quantification across the three scoring dimensions. (A high ATS indicates that a task is being underwritten by proper tool usage.)

% Task Completion is scored on a 10 to 100% scale. This measure is critical if you want to understand how completely each specific, major task is being done.

Color coding is intended to illustrate the nature of the risk accrual for each major task within this phase of the process. Color code risk definitions are found in Chapter 2.

Tools, Methods, and Best Practices That Enable the Analyze Tasks

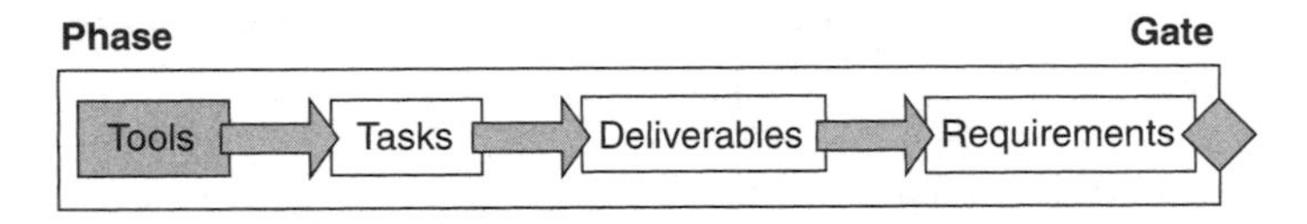

The tool set that supports the Analyze tasks to provide the Analyze deliverables includes the following software, methods, and best practices:

- **Critical Parameter Management** for marketing; marketing requirements and metrics.

TABLE 5.5 Sample Analyze Phase Task Scorecard

1	2	3	4	5	6
SSFM Task	**Average Tool Score**	**% Task Completion**	**Task Results Versus Gate Requirements**	**Risk Color (R, Y, G)**	**Gate Deliverables**
Update marketing critical parameter database					Updated marketing Critical Parameter Management database and product requirements document
Conduct conjoint studies; conduct Monte Carlo simulations; refine price models					Conjoint analysis and Monte Carlo Simulation results; customer documented preferences for form, fit, functions, and features at price points; documented price model
Refine sales forecasts; conduct Monte Carlo simulations					Documented sales forecasts (Monte Carlo simulations)
Define value proposition; define brand positioning strategy (from market innovation strategy); develop alignment between brand positioning strategy and the value proposition					Documented formal product value proposition; documented brand positioning strategy for the value proposition; document alignment of brand positioning strategy to the marketing innovation strategy for the BU
Develop marketing and sales channel strategy					Documented marketing and sales channel strategy
Show marketing objectives aligned with organizational goals					Documented alignment of offering with organizational goals
Update RWW Analysis					Documented updates to the RWW Analysis
Update the business case					Documented updates to the business case
Develop Plan phase project management plan and risk mitigation plan					Documented Plan phase organizational capabilities assessment, project management plan, and risk mitigation plan

- **Customer-based concept testing.**
- **Conjoint studies and analysis.**
- **Design of Experiments (DOE)**—a full and fractional factorial design and sequential experimentation.
- **Descriptive and inferential statistical data analysis.**
- Analysis of Variance (**ANOVA**) data analysis.
- **Regression and empirical modeling** methods.
- **Multivariable studies**—critical marketing parameter screening studies.
- **Statistical Process Control** for marketing data and processes.
- **Price modeling**: Market perceived price profile.
- **Sales forecasting.**
- **Monte Carlo Simulation.**
- **Sales channel analysis.**
- **Brand development and management.**
- **RWW Analysis.**
- **Business case analysis.**
- **Project management tools**, including key events timeline, Monte Carlo Simulation for statistical cycle-time design and forecasting, and FMEA for cycle-time risk assessment.

The sample scorecard, as shown in Table 5.6, is used by marketing teams who apply tools to complete tasks.

Columns 1 and 7 align the tool to a specific task and its requirement for justifying the use of the tool. (Never use a tool if you can't justify its ability to fulfill a task.)

Quality of Tool Usage, Data Integrity, and Results Versus Requirements are scored on a scale of 1 to 10 using the simple scoring templates provided in Chapter 2.

TABLE 5.6 Sample Analyze Phase Tools Scorecard

1	2	3	4	5	6	7
SSFM Tool	**Quality of Tool Usage**	**Data Integrity**	**Results Versus Requirements**	**Average Tool Score**	**Data Summary, Including Type and Units**	**Task**
Critical Parameter Management methods; Statistical Process Control, capability studies, DOE						Update marketing critical parameter database
Conjoint analysis; DOE; descriptive and inferential statistical data analysis, multivariable studies, ANOVA, regression customer-based concept testing; Monte Carlo Simulation; price modeling						Conduct conjoint studies; conduct Monte Carlo simulations; refine price models
Sales forecasting methods; Monte Carlo simulation						Refine sales forecasts; conduct Monte Carlo Simulations
Customer value management tools; brand management methods						Define value proposition; define brand positioning strategy (from market innovation strategy); develop alignment between brand positioning strategy and the value proposition
Channel mapping tools						Develop marketing and sales channel strategy
GOSPA process/S.M.A.R.T.						Show marketing objectives aligned with company's strategic goals, division, product management
RWW Analysis methods						Update RWW Analysis
Business case development methods						Update the business case
Project management methods, Monte Carlo simulation; project FMEA						Develop Plan phase project management plan and risk mitigation plan

Average Tool Score illustrates overall tool quantification across the three scoring dimensions. (A high ATS indicates that a task is being underwritten by proper tool usage.)

Data Summary is intended to illustrate the nature of the data and the specific units of measure (attribute or continuous data, scale or specific units of measure).

The Plan Phase

Working in parallel with and in concert with its technical counterparts, during the Plan phase of UAPL, the marketing team focuses on developing detailed, robust marketing and sales plans to launch the product into the outbound marketing, sales, and customer support environments. Marketing ensures that the detailed elements (critical tasks and their measurable results) of the plans have direct measures and linkages to being able to adjust mean performance to targets for reaching the entitlement of the business case. Channel definition, development, and management plans are conducted within this phase.

Requirements for the Plan Gate

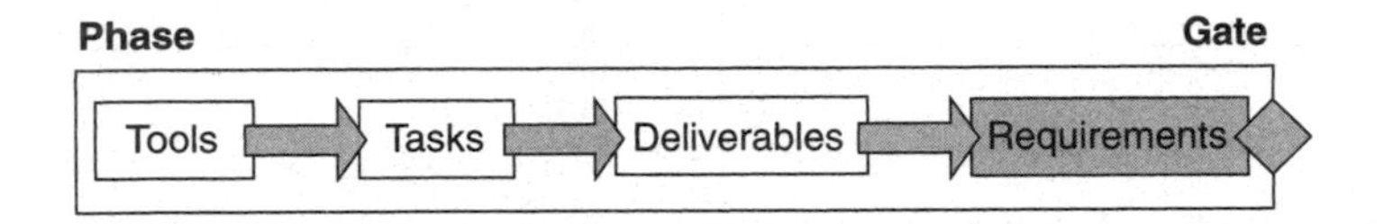

The third phase of UAPL requirements builds on the work completed in the two prior phases. The Plan phase has eight requirements:

1. Robust marketing plan developed
2. Robust sales channel plan created

3. Marketing collateral, advertising, and promotional materials developed
4. Channel management plan documented
5. Marketing and sales critical parameters updated and respective database updated
6. RWW Analysis refined and updated
7. Business case refined and updated
8. Launch phase project management plan developed

Plan Phase Tools, Tasks, and Deliverables

Producing the Plan phase deliverables requires an investment in numerous, detailed inbound marketing tasks enabled by specific tools, methods, and best practices. Their integration ensures that the right data is being developed and summarized to fulfill the major requirements at the Plan gate review (see Figure 5.7).

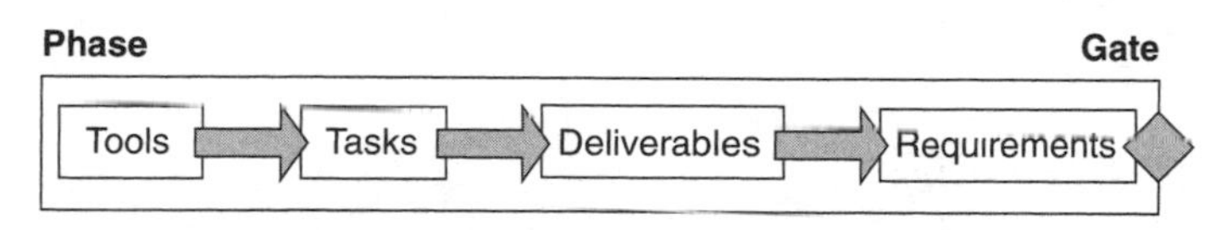

FIGURE 5.7 The Plan gate review.

The Plan phase is the critical foundation for the next Launch step of the UAPL process and the activities and deliverables for the operational process to follow. Planning tasks are easy to gloss over. However, marketing teams that invest time in thorough planning will reap the benefits of doing it right the first time and will avoid significant rework. The tools enable the tasks to produce the deliverables that meet the phase-gate requirements and ensure measurable results. The following list summarizes the set of tools, tasks, and deliverables for the Plan phase (see Figure 5.8).

Tools

- Marketing and sales requirements, metrics, and Critical Parameter Management
- Marketing and sales management plan development methods
- Marketing and sales channel process mapping and analysis
- Marketing and sales process noise diagramming
- Brand positioning and management
- Marketing communications planning
- Marketing collateral and promotional material planning and development
- Customer relationship management methods
- RWW Analysis
- Business case analysis
- The hybrid grid
- Project management tools

Tasks

- Audit marketing plan
- Create and implement internal communications plan
- Develop external communications strategy
- Develop sales support plan
- Develop pricing plan by segment
- Update business case and sales forecast
- Update marketing plan
- Document critical risks, problems, and assumptions
- Implement Critical Parameter Management database
- Develop who/what matrix to CTQs
- Document process maps depicting how to fulfill CTQ requirements
- Establish market accountability plan
- Test effectiveness of product positioning
- Create LAUNCH phase marketing and sales project plan

Deliverables

- Updated Critical Parameter Management database for marketing
- Initial Critical Parameter Management database for sales
- Process maps for marketing and sales processes
- Plans: preliminary marketing, sales, channel management, market communications, customer relationship management and brand positioning plans
- Preliminary designs for marketing and sales collateral, advertising, and promotional materials
- Refined and updated RWW Analysis
- Refined and updated business case
- LAUNCH phase project management plans (Monte Carlo simulations)
- Documented LAUNCH phase project plans

FIGURE 5.8 The tools-tasks-deliverables for the Plan phase.

Plan Gate Deliverables

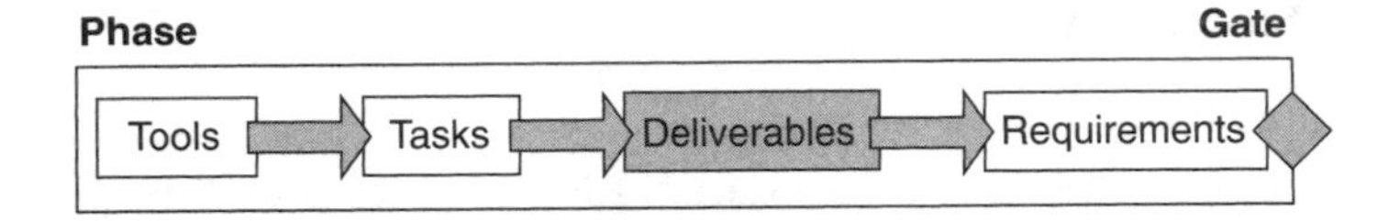

The deliverables at the Plan gate are all major deliverables. They are critical outputs to this phase, and they should be inspected for completeness and accuracy. The many major deliverables are all equally important; no one output should be overlooked or incomplete. The eight Plan deliverables include the following:

1. Updated **Critical Parameter Management database for marketing.**

2. **Initial Critical Parameter Management database for sales.**

3. **Process maps for marketing and sales** processes.

4. **Plans:** preliminary marketing, sales channel management, market communications, customer relationship management, and brand positioning.

5. **Preliminary designs for marketing and sales collateral, advertising, and promotional materials.**

6. Refined and updated **RWW Analysis**.

7. Refined and updated **business case.**

8. **Launch phase project management plans** (Monte Carlo simulations).

The sample scorecard, as shown in Table 5.7, is used by marketing gatekeepers who manage risk and make functional gate decisions for a specific project as part of the portfolio of projects being conducted by the marketing organization.

Columns 1 and 6 align the gate deliverable to a gate requirement. Each deliverable is justified as it contributes to meeting a gate requirement. (Never produce a deliverable if you can't justify its ability to fulfill a gate requirement.)

Grand Average Tool Score (GATS) illustrates aggregated tool quantification across the three scoring dimensions. (A high GATS indicates that a group of tasks is underwriting a gate deliverable.)

% Task Completion is scored on a 10 to 100% scale. This measure is critical if you want to understand how completely a group of related tasks are fulfilling a major gate deliverable.

Color coding is intended to illustrate the nature of the risk accrual for each major deliverable within this phase of the process. Color code risk definitions are found in Chapter 2.

TABLE 5.7 Sample Plan Gate Deliverables Scorecard

1	2	3	4	5	6
SSFM Gate Deliverable	**Grand Average Tool Score**	**% Task Completion**	**Task Results Versus Gate Deliverable Requirements**	**Risk Color (R, Y, G)**	**Gate Requirement**
Process maps for marketing and sales processes					Develop robust marketing and sales plans
Documented robust marketing plan					Develop a robust marketing plan
Documented robust sales channel plan					Develop a robust sales channel plan
Marketing collateral, advertising, and promotional materials					Develop marketing collateral, advertising, and promotional materials
Documented CRM plan					Document a CRM plan
Documented marcom plan					Document a marcom plan
Documented brand positioning plan					Document a brand positioning plan
Documented channel management and distribution plans					Document a channel management plan; document a distribution plan
Documented updates to marketing and sales critical parameters database					Update marketing and sales critical parameters database
Generate initial sales critical parameters database					Document sales critical parameters and database
Documented updates to RWW Analysis					Refine and update RWW Analysis
Documented update to the business case					Document updated business case
Documented Launch phase project management plan and risk mitigation plan					Document Launch phase project management plan and risk mitigation plan

Plan Phase Tasks

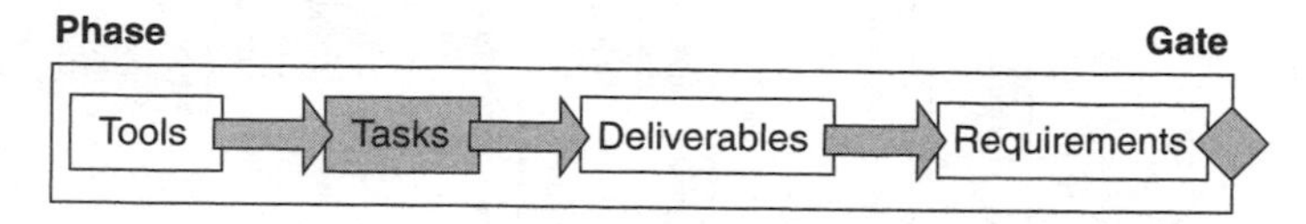

Dictated by the Plan deliverables required to complete the third UAPL phase, the tasks within the Plan phase include the following:

1. Audit the marketing plan.
2. Create and implement an internal communications plan for the company employees and across the value chain, including the sales force, customer support, services, and suppliers.
3. Develop an external communications strategy (including multiple venues and format options such as collaterals and the Internet/Web).
4. Develop a sales support plan.
5. Develop a pricing plan by segment.
6. Update the business case and sales forecast.
7. Update the marketing plan.
8. Document critical risks, problems, and assumptions.
9. Implement a Critical Parameter Management database.
10. Develop who/what matrices to map responsibility for delivery of all CTQs.
11. Document process maps depicting how to fulfill the CTQ requirements.
12. Establish a market accountability plan.
13. Test effectiveness of product positioning.
14. Create Launch phase marketing and sales project plan and risk analysis.

The sample scorecard, as shown in Table 5.8, is used by marketing project team leaders who manage major tasks and their time lines as part of their project management responsibilities.

TABLE 5.8 Sample Plan Phase Task Scorecard

1	2	3	4	5	6
SSFM Task	Average Tool Score	% Task Completion	Task Results Versus Gate Requirements	Risk Color (R, Y, G)	Gate Deliverable
Conduct market channel, distribution and sales process mapping and noise diagramming; conduct current market trend analysis					Documented marketing and sales process maps
Define customer relationship management plan					Documented customer relationship management plan
Develop marcom plan					Documented marcom plan
Define product or service positioning relative to brand strategy					Documented brand positioning plan
Develop channel management and distribution plan					Documented channel management and distribution plan
Develop robust marketing plan					Documented robust marketing plan
Develop robust sales channel plan					Documented robust sales plan
Develop marketing collateral, advertising, and promotional materials					Marketing collateral, advertising, and promotional materials
Develop channel management and distribution plan					Documented channel management and distribution plan
Update marketing critical parameters database					Documented updates to marketing and sales critical parameters database
Develop sales critical parameters database					Documented sales critical parameters database
Update RWW Analysis					Documented updates to RWW Analysis
Update business case					Documented update to the business case
Create Launch phase project management plan and risk mitigation plan					Documented Launch phase project management plan and risk mitigation plan

Columns 1 and 6 align the task to a specific gate deliverable for justifying the task. (Never conduct a task if you can't justify its ability to produce a gate deliverable and fulfill a gate requirement.)

Average Tool Score (ATS) illustrates overall tool quantification across the three scoring dimensions. (A high ATS indicates that a task is being underwritten by proper tool usage.)

% Task Completion is scored on a 10 to 100% scale. This measure is critical if you want to understand how completely each specific, major task is being done.

Color coding is intended to illustrate the nature of the risk accrual for each major task within this phase of the process. Color code risk definitions are found in Chapter 2.

Tools, Methods, and Best Practices That Enable the Plan Tasks

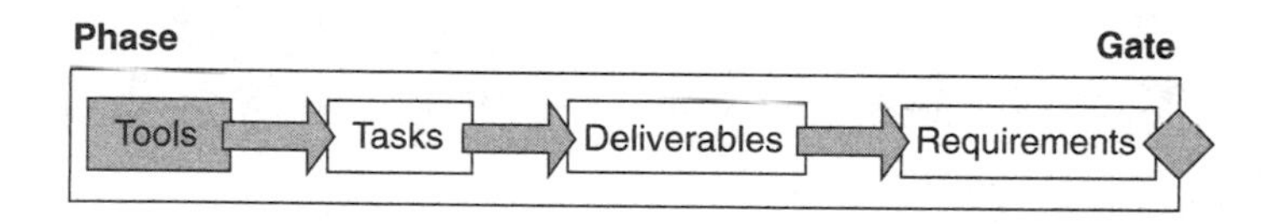

The tool set that supports the Plan tasks to provide the Plan phase deliverables includes the following methods and best practices:

- **Critical Parameter Management** for marketing and sales channel, including requirements and metrics.
- **Marketing and sales channel process mapping** and analysis for market trend analysis, customer behavioral dynamics mapping, competitive position analysis, Won-Lost Analysis, key events time line, SWOT Analysis, Porter's 5 Forces Analysis, and market perceived quality profile.
- **Marketing and sales plan development methods.**

- **Marketing and sales process noise diagramming.**
- **Customer relationship management methods.**
- **Marketing communications planning.**
- **Brand positioning and management.**
- **Marketing collateral and promotional material planning and development.**
- **RWW Analysis.**
- **Business case analysis.**
- **The hybrid grid.**
- **Project management tools,** including key events time line, Monte Carlo Simulation for statistical cycle-time design and forecasting, and FMEA for cycle-time risk assessment.

The sample scorecard, as shown in Table 5.9, is used by marketing teams who apply tools to complete tasks.

Columns 1 and 7 align the tool to a specific task and its requirement for justifying the use of the tool. (Never use a tool if you can't justify its ability to fulfill a task.)

Quality of Tool Usage, Data Integrity, and Results Versus Requirements are scored on a scale of 1 to 10 using the simple scoring templates provided in Chapter 2.

Average Tool Score illustrates overall tool quantification across the three scoring dimensions. (A high ATS indicates that a task is being underwritten by proper tool usage.)

Data Summary is intended to illustrate the nature of the data and the specific units of measure (attribute or continuous data, scale or specific units of measure).

TABLE 5.9 Sample Plan Phase Tools Scorecard

1	2	3	4	5	6	7
SSFM Tool	**Quality of Tool Usage**	**Data Integrity**	**Results Versus Requirements**	**Average Tool Score**	**Data Summary, Including Type and Units**	**Task**
Process mapping tools						Conduct market channel, distribution, and sales process mapping
Customer behavioral dynamics mapping; competitive position analysis; Won-Lost Analysis; key events time line; SWOT Analysis; Porter's 5 Forces Analysis; market perceived quality profile						Conduct current market trend analysis
Marketing plan development methods; marketing process noise diagramming						Develop robust marketing plan
Sales plan development methods; sales process noise diagramming						Develop robust sales channel plan
Customer relationship management methods						Define customer relationship management plan
Marketing communications planning						Develop marcom plan

(*continues*)

TABLE 5.9 Sample Plan Phase Tools Scorecard (continued)

1	2	3	4	5	6	7
SSFM Tool	Quality of Tool Usage	Data Integrity	Results Versus Requirements	Average Tool Score	Data Summary, Including Type and Units	Task
Process mapping tools						Conduct market channel, distribution, and sales process mapping
Brand positioning and management tools						Define product or service positioning relative to brand strategy
Marketing collateral and promotional material planning and development methods						Develop marketing collateral, advertising, and promotional materials
Critical Parameter Management methods						Update marketing critical parameters database
Critical Parameter Management methods						Develop sales critical parameters database
Channel management tools; distribution planning tools						Develop channel management and distribution plan
RWW Analysis methods						Update RWW Analysis
Business case development methods						Update business case
Project management methods, Monte Carlo simulation; project FMEA						Develop Launch phase project management plan and risk mitigation plan

The Launch Phase

This final phase of the UAPL tactical process involves contributions from multiple disciplines throughout the company. The commercialization team puts its final touches on things to prepare the offering for the marketplace. During the Launch phase, marketing focuses on developing a detailed, measurable launch process (with key measurable inputs and outputs) to ensure that the launch parameters can be tuned and are robust to sources of variation in the launch environment. Marketing documents and implements the launch process control plan that measures performance and enables a proactive (or reactive) response to incoming data.

Leading and lagging indicators of launch dynamics and performance are developed and in place to control the launch process against its targets. Marketing should scrutinize the launch mechanics and the system of critical parameters to ensure that the right data is gathered, in the appropriate units of measure, to enable rapid adjustments to be made to meet launch targets even in the presence of launch noise. Marketing should use Statistical Process Control to track trends in key launch functions.

Requirements for the Launch Gate

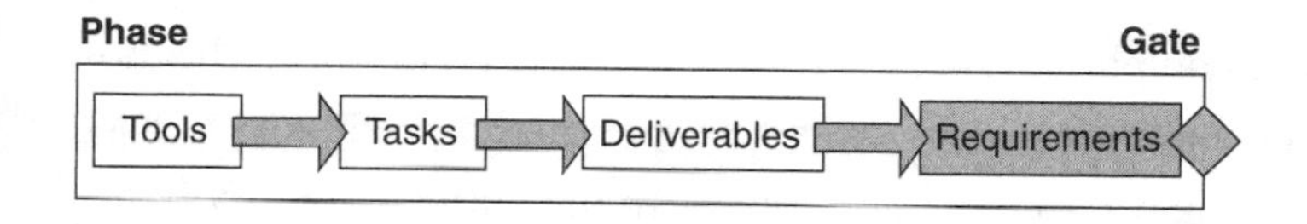

1. Documented launch plan.
2. Documented post-launch marketing and sales channel processes and control plans.
3. Documented customer relationship management process and control plan.

4. Final marketing and sales collaterals, advertising, and promotional materials.
5. Updated marketing and sales-critical parameters and database.
6. Refined and updated RWW Analysis.
7. Refined and updated business case and sales forecast.

Launch Phase Tools, Tasks, and Deliverables

Producing the Launch phase deliverables requires an investment in numerous detailed inbound and outbound marketing tasks enabled by specific tools, methods, and best practices. Their integration ensures that the right data is being developed and summarized to fulfill the major requirements at the Launch gate review (see Figure 5.9).

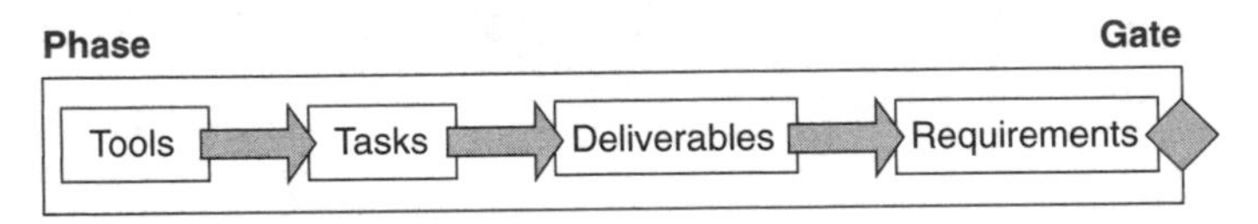

FIGURE 5.9 The Launch gate review.

The Launch phase, the final step of the UAPL commercialization process, is when the window dressing is put on the product or service before it is released to the marketplace. In the anticipation of quickly getting the new offering to market, too often the excitement blinds the commercialization team such that key activities and deliverables get rushed and are launched as incomplete. The following set of tools, tasks, and deliverables enables the Launch phase-gate requirements to be met and ensure measurable results (see Figure 5.10).

Tools

- Critical Parameter Management for marketing and sales channel requirements and metrics
- Launch planning methods
- Control plan development methods
- Marketing and sales channel process mapping and analysis
- Marketing and sales process noise diagramming
- Marketing plan audit
- Statistical Process Control, statistical data mining, and analysis tools
- Customer relationship management methods
- Marketing communications planning
- Brand positioning and management
- RWW Analysis
- Business case analysis
- Project management tools

Tasks

- Conduct marketing plan audit
- Assess internal communications plan
- Assess implementation plan readiness

Deliverables

- Integrated value chain plan
- Assessment of sales and support risk
- Launch plan
- Control plans and process maps of post-launch processes
- Statistical Process Control charts or critical sales and marketing data
- Marketing and sales process noise diagrams and FMEA
- Critical Parameter Management databases for marketing and sales requirements, metrics, and controls
- Final marketing and sales collaterals, advertising, and promotional materials
- Final documentation for brand alignment with value proposition
- Updated and refined RWW Analysis
- Updated and refined business case

FIGURE 5.10 The tools-tasks-deliverables for the Launch phase.

Launch Gate Deliverables

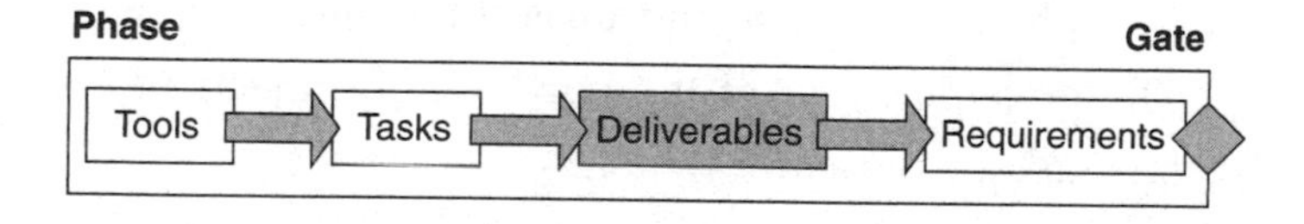

The deliverables for the Launch phase-gate are all major deliverables. They are critical elements to finalizing the commercialization of a product or service, and they should be inspected for completeness and accuracy. The major deliverables are equally important and should not be shortchanged. The 11 Launch deliverables are as follows:

- **Integrated value chain plan** (including product/services, production, supply chain, sales, service, and support).
- **Assessment of sales and support risk.**
- **Launch plan.**

- **Control plans and process maps of post-launch processes** (marketing, sales channel management, customer relationship management, customer support, supporting alliances, and suppliers management).
- **Statistical Process Control charts** on critical sales and marketing data.
- **Marketing and sales process noise diagrams and FMEA.**
- **Critical Parameter Management databases** for marketing and sales requirements, metrics, and controls.
- **Final marketing and sales collaterals, advertising, and promotional materials.**
- **Final documentation for brand alignment with value proposition.**
- Updated and refined **RWW Analysis.**
- Updated and refined **business case and sales forecast.**

The sample scorecard, as shown in Table 5.10, is used by marketing gatekeepers who manage risk and make functional gate decisions for a specific project as part of the portfolio of projects being conducted by the marketing organization.

Columns 1 and 6 align the gate deliverable to a gate requirement. Each deliverable is justified as it contributes to meeting a gate requirement. (Never produce a deliverable if you can't justify its ability to fulfill a gate requirement.)

Grand Average Tool Score (GATS) illustrates aggregated tool quantification across the three scoring dimensions. (A high GATS indicates that a group of tasks is underwriting a gate deliverable.)

% Task Completion is scored on a 10 to 100% scale. This measure is critical if you want to understand how completely a group of related tasks are fulfilling a major gate deliverable.

Color coding is intended to illustrate the nature of the risk accrual for each major deliverable within this phase of the process. Color code risk definitions are found in Chapter 2.

TABLE 5.10 Sample Launch Gate Deliverables Scorecard

1	2	3	4	5	6
SSFM Gate Deliverable	**Grand Average Tool Score**	**% Task Completion**	**Task Results Versus Gate Deliverable Requirements**	**Risk Color (R, Y, G)**	**Gate Requirements**
Documented launch plan					Show detailed product service, production, supply chain, sales, service, and support plans detail and integration. Show how awareness, consideration, trial, usage, and referral will be promoted and measured.
Documented post-launch marketing and sales channel processes and control plans					Control plans and process maps of marketing, sales channel, customer service, support, and other Critical to Quality functions
Documented customer relationship management process and control plan					Control plans and process maps of customer relationship management
Final marketing and sales collateral, advertising and promotional materials plan					Show how activities align with marketing objectives. Show deployment schedule: what will be done, when, and at what cost.
Updated marketing and sales critical parameter database					Statistical Process Control charts on critical sales and marketing data
Refined, updated RWW Analysis					Offer latest assessment of risk and potential revenues against last iteration of product. Show FMEA.
Refined, updated business case sales forecast					Show how value proposition aligns and supports branding strategy

Launch Phase Tasks

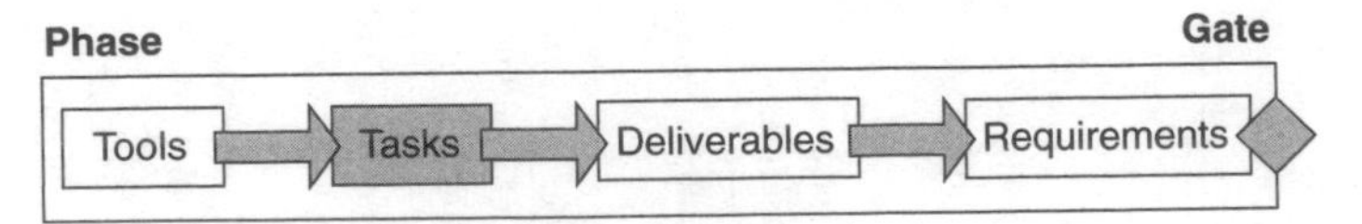

Dictated by the Launch deliverables required to complete the final UAPL phase, the short list of tasks within the Launch phase is as follows:

1. Conduct marketing plan audit.
2. Assess the internal communications plan (employees and value chain players such as the sales force, customer support and service, partners, suppliers).
3. Assess implementation plan readiness of the external communications strategy implementation, sales support plan, channel support plan, sales promotion plan, public relations plan.

The sample scorecard, as shown in Table 5.11, is used by marketing project team leaders who manage major tasks and their time lines as part of their project management responsibilities.

Columns 1 and 6 align the task to a specific gate deliverable for justifying the task. (Never conduct a task if you can't justify its ability to produce a gate deliverable and fulfill a gate requirement.)

Average Tool Score (ATS) illustrates overall tool quantification across the three scoring dimensions. (A high ATS indicates that a task is being underwritten by proper tool usage.)

% Task Completion is scored on a 10 to 100% scale. This measure is critical if you want to understand how completely each specific, major task is being done.

Color coding is intended to illustrate the nature of the risk accrual for each major task within this phase of the process. Color code risk definitions are found in Chapter 2.

TABLE 5.11 Sample Launch Phase Task Scorecard

1	2	3	4	5	6
SSFM Task	Average Tool Score	% Task Completion	Task Results Versus Gate Requirements	Risk Color (R, Y, G)	Gate Deliverable
Show detailed product service, production, supply chain, sales, service support plan integration					Documented launch plan
Show plan to promote and measure awareness, consideration, trial, usage, and referral					Documented launch plan
Develop process maps and control plans for marketing, sales channel, customer service, support, and other Critical to Quality functions					Documented post-launch marketing and sales channel processes and control plans
Develop process maps and control plans for customer relationship management					Documented customer relationship management process/control plan
Show how activities align with marketing objectives. Develop deployment schedule. Show costs.					Final marketing and sales collateral, advertising, and promotional materials plan
Show Statistical Process Control charts on critical marketing and sales data.					Updated marketing and sales critical parameter database
Show latest assessment of risk and potential revenues against latest product iteration. Show FMEA.					Refined, updated RWW Analysis
Develop final sales forecast with pricing strategy and integrate into business case					Refined, updated business case sales forecast

Tools, Methods, and Best Practices That Enable the Launch Tasks

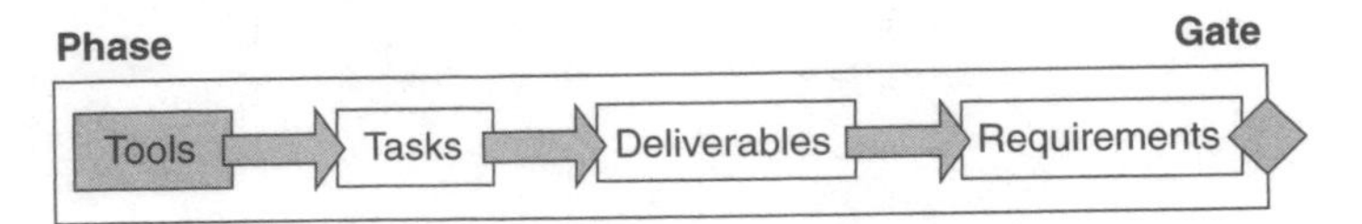

The tool set that supports the Launch tasks to provide the Launch phase-gate deliverables includes the following methods and best practices:

- **Critical Parameter Management** for marketing and the sales channel requirements and metrics.
- **Launch planning methods.**
- **Control plan development methods.**
- **Marketing and sales channel process mapping** and analysis.
- Marketing and sales **process noise diagramming.**
- **Marketing plan audit.**
- **Statistical Process Control, statistical data mining, and analysis tools.**
- **Customer relationship management methods.**
- **Marketing communications planning.**
- **Brand positioning and management** (including development methods for marketing collaterals and promotional materials).
- **RWW Analysis.**
- **Business case analysis.**
- **Project management tools**, including key events time line, Monte Carlo Simulation for statistical cycle-time design and forecasting, and FMEA for cycle-time risk assessment.

The sample scorecard, as shown in Table 5.12, is used by marketing teams who apply tools to complete tasks.

TABLE 5.12 Sample Launch Phase Tools Scorecard

1	2	3	4	5	6	7
SSFM Tool	**Quality of Tool Usage**	**Data Integrity**	**Results Versus Requirements**	**Average Tool Score**	**Data Summary, Including Type and Units**	**Task**
Process mapping: swim lane flowcharts						Show detailed product service, production, supply chain, sales, service support plan integration
Program management methods/ RACI matrix						Show plan to promote and measure awareness, consideration, trial, usage, and referral
Process mapping: swim lane flowcharts/ RACI matrix/who-what matrix; key events time line, Monte Carlo Simulation Critical Parameter Management						Develop process maps and control plans for marketing, sales channel, customer service, support, and other Critical to Quality functions
Process mapping: swim lane flowcharts/ RACI matrix/who-what matrix						Develop process maps and control plans for customer relationship management
GOSPA method/RACI matrix/ program management methods, Critical Parameter Management						Show alignment of activities to marketing objectives. Develop deployment schedule. Show costs.
SPC methods Critical Parameter Management						Show Statistical Process Control charts on critical marketing and sales data
MPQP gap analysis/FMEA/technical assessment/SWOT/Porter's 5 Forces Analysis						Show latest assessment of risk and potential revenues against latest product iteration. Show FMEA.
NPV/IRR/develop sales forecast based on planned and stressed product mix strategy						Develop final sales forecast with pricing strategy and integrate into business case

Columns 1 and 7 align the tool to a specific task and its requirement for justifying the use of the tool. (Never use a tool if you can't justify its ability to fulfill a task.)

Quality of Tool Usage, Data Integrity, and Results Versus Requirements are scored on a scale of 1 to 10 using the simple scoring templates provided in Chapter 2.

Average Tool Score illustrates overall tool quantification across the three scoring dimensions. (A high ATS indicates that a task is being underwritten by proper tool usage.)

Data Summary is intended to illustrate the nature of the data and the specific units of measure (attribute or continuous data, scale or specific units of measure).

An area of curiosity with many of our clients is what constitutes a "rigorously defined launch control plan with appropriate measures of marketing and sales performance." Although the answer is situational, depending on your business, value chain, and current processes, we recommend some basic core elements for each implementation. Let's examine control plan and launch plan separately.

A control plan ensures that a commercialization project's recommendations become part of the business's ongoing operations. It identifies an integrated system of people, process, and technology needed to

- Sustain the performance improvements
- Drive continuous improvement
- Proactively prevent problems

Every control plan is unique, but it should contain the minimum core elements: a control method (the control charts, specific metrics, and audit process) and a risk management and response plan.

A launch plan defines *how to transition* the commercialization project to ongoing operations—the process owner(s) and the process "players" as a part of their day-to-day activities. It is a higher-level document that encompasses the control plan. A launch plan should include the following:

- A transition plan from the commercialization project team to the value chain
- (Updated) procedural documentation (as appropriate)
- A training plan
- (Updated) roles and responsibilities documentation (RACI)
- Updated financials
- A communications plan (internal and external)
- Any lessons learned or special conditions to share with operations
- The control plan

The commercialization team should work out any kinks before transferring responsibility to operations. A well-developed transition plan often evolves as a collaborative effort between the commercialization project and key process owners throughout the value chain. During the next process area (operations), the commercialization team should check to ensure that operations are running smoothly, according to plan, before closing out their commercialization process. However, Murphy's Law rules: Don't expect perfection; thorough planning will help launch and operations be flexible and adaptable.

Summary

You now have enhanced your product and services commercialization process with a good foundation of Six Sigma for inbound marketing. The Understand-Analyze-Plan-Launch approach provides strong linkage to the technical departments in your firm, such that both disciplines now have a common language to communicate and translate customer requirements into business and product (or services requirements). Plus, outputs from this process lead you into the ongoing operational process of managing your launched offering.

Recall that it is essential not to rush the tactical process of product commercialization. Taking the time to *do things right the first time*, avoiding rework, in the long run will save the firm money and improve customer satisfaction. To unleash the power of Six Sigma—to manage by fact—you need to ensure that your inbound marketing activities adhere to an established process, such that the commercialization team clearly understands its requirements and the tools, tasks, and deliverables needed to satisfy those requirements. UAPL provides you that structure to ensure measurable and predictable results that fulfill the phase-gate requirements. You may choose to call your process phases by different names—that's fine. *What you do and what you measure* are what really matter.

Now that you understand the inbound marketing processes of portfolio renewal and commercialization, let's move on to outbound marketing and examine how Six Sigma can support marketing and sales in the operational process of managing a launched product or service.

6

Six Sigma in the Operational Marketing Process

Outbound Marketing for Post-Launch Line Management and Sales for Products and Services

Six Sigma in the Operational Marketing Process

The commercialization team has completed its preparation to launch a new offering. Its deliverables have met the process requirements and have passed the final launch phase-gate review. The transition from inbound to outbound marketing marks the start of this next process area—operations. Marketing now may execute the plans developed during the commercialization process—including the launch plan, control plans, communication plans, marketing communications (marcom) plans, and training plans. This activity triggers outbound marketing's go-to-market partners to incorporate the new offering into their day-to-day processes: sales channel distribution, services, customer support, public relations, and administration (order-to-collection).

The market's perception of your firm's value forms from experience with your offerings and communications. Presenting a united front along the value chain and aligning and integrating both your tangible and intangible assets constructs a powerful proposition. The culmination of this work occurs in this Operational process arena. How well you execute the prior commercialization process and manage your go-to-market resources often determines your enterprise's success—both financially and with your brand equity. The commercialization process, if done well, provides the planning to execute right the first time, with few to no recalls or rework in the field. Sales data will tell you if the mix of product, promotion, and place justifies the asking price. Missed sales targets, product recalls and patches, and high customer support costs often result from poor commercialization planning. Poorly communicated expectations and the critical parameters to be managed can cause your value chain partners to incur waste in cycle time and cost. Mismanaged go-to-market resources alone also can be a significant drain on a business. When the eventual marketplace change (or shift) happens, it only compounds the inefficiencies of poor planning and a poorly managed operational team.

The Six Sigma for Growth discipline can strengthen your current management approaches and provide tools to give your field management better information so that they can make better decisions. Although this chapter suggests a unique operational set of tools-tasks-deliverables to add to your current marketing and sales arsenal, these should supplement and bolster your current methods, tools, and best practices.

While conducting the work required for meeting sales and market share objectives, it is easy to get caught up in a sea of unexpected variation. Plans are made, and then suddenly everything changes. Teams can go off-plan fast. With Six Sigma, this means that assignable causes of variation (noise) cause the marketing and sales processes to go out of control. This is one reason why some marketing and sales professionals may choose to restrain their investment in developing detailed launch plans—they know how quickly their plan has to change. It is tempting to react to whatever comes along. Significant levels of planning might seem like a waste of time. More often than not, something triggers a change in the plan, so why bother planning? Becoming good at reacting to change is *not* the end goal. Aim to focus on the vital few (the verified critical) parameters and proactive preventive plans to keep your operations on target to sustain growth.

Change, in the operational environment of post-launch marketing and sales, is normal. If you are surprised by it, you probably haven't been doing product launches or managing field operations for very long. To be good at sustaining growth, you have to anticipate changes by learning to measure things that help your team adjust the flow and nature of your marketing and sales tasks. You want to anticipate positive and negative variation within your marketing and sales processes so that your team can make adjustments that result in the desired outcomes. You need to be able to design a plan that is adaptive by design so that you rarely have to go off-plan.

What does it mean to anticipate change? It means you have to measure leading indicators of market, competitive, and customer behavioral dynamics that prepare your team to make preplanned changes in your tasks and tools, *not your plan*! You can frame this proactive measurement approach by stating, "Measure the probability of impending failures."

A simple analogy helps illustrate this point. Heart attacks are a leading cause of death in the adult population (a "failure" you can count). Obesity (body mass index) is a measurable attribute that is a proven link to causing a heart attack (a "quality attribute" you can measure). Too much food and too little exercise are measurable attributes linked to the root causes of obesity (gross fundamental measures of causes of obesity). The specific kinds of food one eats and how much and the time and type of exercise one gets are basic, fundamental measures. They lie at the root of controllable, adjustable factors that can prevent the onset of obesity and ultimately control the risk of a heart attack. If you carefully plan a diet-and-exercise regime, you can adaptively control your lifestyle to lower your risk of a heart attack, even if you are genetically predisposed to heart problems. Of course, this requires discipline. You have to measure the fundamentals that directly cause a functional, traceable dynamic that leads to "failure." Obesity and heart attacks are *lagging indicators*. You can react to them only after they are detected. Food and exercise are *leading indicators* that can tell you when to make changes to help avoid obesity and heart attacks.

Don't wait for lagging indicators to scare you into action once they have developed to the point where you can detect them. Measuring the fundamentals of human behavior, decision-making, and purchasing behavior will set you on a course of action that lets you prevent marketing and sales process failures. To become good at this approach, you have to be a student of post-launch marketing and sales "noises."

A noise is any source of variation. They often precede an outright failure in a process. They are changes that affect the mechanics—the fundamental cause-and-effect relationships inside marketing and sales processes. Processes contain tasks or steps that you carry out to produce a targeted result. Noises alter the targeted results you are attempting to achieve. They disrupt the fundamental functions that control results. There are three general sources of noise:

- **External variation**—noise coming from outside the system or process, such as competition, politics, economy, or environment
- **Unit-to-unit variation**—the ability to reproduce a unit or process the same way each time, often due to either production or assembly (a type of internal noise)
- **Deterioration noise**—caused by wearing down or degrading a product or process over time (another type of inner noise)

If a sale didn't close, perhaps an attractive competitive offering was introduced that day (an external variation). Perhaps a computer virus infected your systems such that the sales proposal process took too long to approve an exception to the terms and conditions of a major new sales contract, and the customer couldn't wait any longer (an external variation). Perhaps a salesperson failed to mention appropriate and critical benefits to the prospective customers (a unit-to-unit variation). Or perhaps the marketing collaterals were outdated and failed to address critical customer concerns about Total Cost of Ownership (TCO, a deterioration variation). This chapter explores how to apply Six Sigma discipline in the presence of noise during the launch, management, adaptation, and final discontinuance so that you know when to make the necessary adjustments to stay on plan. Figure 6.1 shows the various kinds of noise on a process or system.

In the Operational marketing process environment, we have identified a generic process segmented into four distinct phases. The

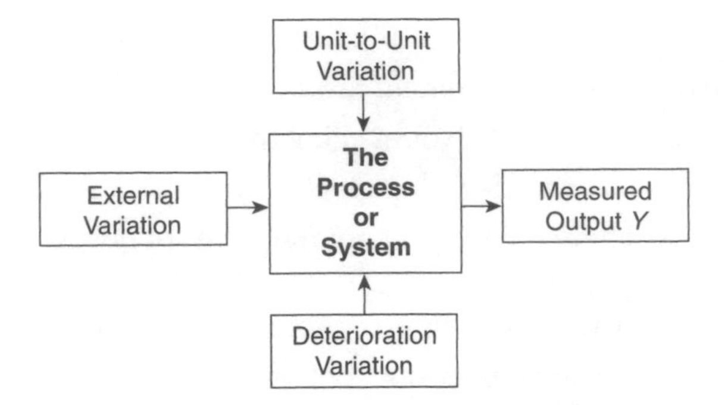

FIGURE 6.1 Assignable causes of variation or noise.

LMAD process controls the life cycle of the portfolio of launched products and services:

1. **Launch** the product or service through its introductory period into the market.
2. **Manage** the offering in the steady-state marketing and sales processes.
3. **Adapt** the marketing and sales tasks and tools as "noises" require change.
4. **Discontinue** the offering with discipline to sustain brand loyalty.

This four-phase approach ebbs and flows in and out of the Manage and Adapt phases to stay on plan throughout an offering's life. Figure 6.2 shows the need for a course correction across a generic LMAD life cycle.

The LMAD process focuses on taking preventive action rather than waiting for a lagging performance indicator to goad the team into chartering a special DMAIC problem-solving project because you failed to make your numbers in the past two quarters. The combination of the four phases establishes a proactive system of critical marketing and sales parameters to limit the number of problems by design. Enabled by preplanned Six Sigma marketing and sales tasks

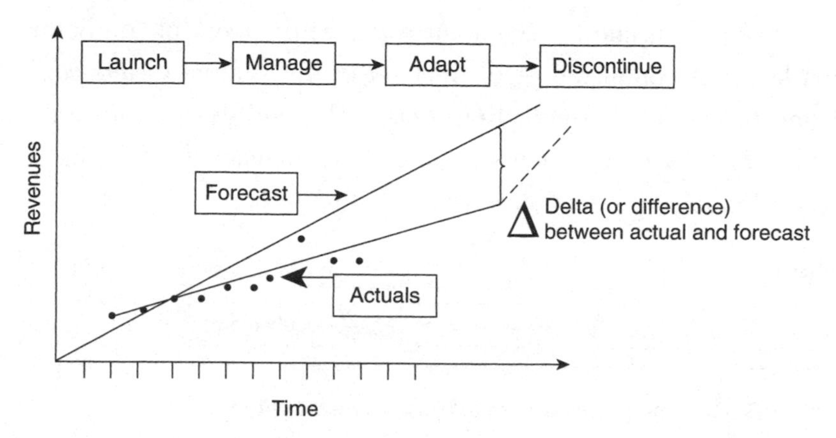

FIGURE 6.2 Generic sales analysis: actual data mapped against the forecast.

and tools to prevent problems, the LMAD phases replace the need for reactive "emergency" DMAIC steps to react to problems. Think of LMAD as a health club versus an emergency room. We prefer to avoid heart attacks.

The phase Gate Review approach applied to the strategic and tactical marketing processes in the preceding two chapters must change now that we are focusing on the post-launch environment. *Operational marketing processes* behave differently than portfolio renewal or commercialization processes. Like steady-state manufacturing, operational environments *lack formal phase gates*. Hence, operations require timely, periodic reviews of progress against the marketing and sales plans. These should be thought of as *key milestones in the continuum* of executing the marketing and sales plan, not as Gate Reviews. Data and results can be gathered and summarized on an hourly, daily, weekly, monthly, quarterly, and yearly basis. It is essential that you balance between proactive and reactive performance measures. Mostly, you want to focus on leading performance indicators if you want to stay on plan and under control. You have to be realistic and recognize that it is almost impossible not to measure some lagging indicators—you just want to keep them to a minimum. This is part of what it means to be "lean."

Post-launch line management is where the work of "outbound" marketing is conducted. A distinct set of measurable results comes from this work. The marketing tools, tasks, and deliverables are all defined by the requirements for launching, managing, adapting, and discontinuing a product or service (see Figure 6.3).

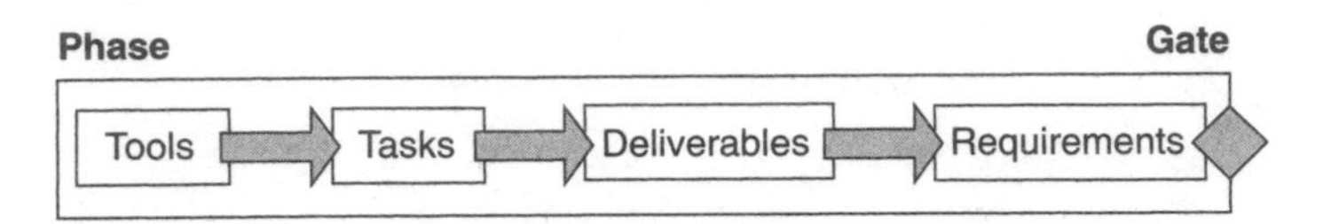

FIGURE 6.3 The phase-key milestone review system.

Most of the tools, methods, and best practices of "outbound" marketing are the same as "inbound" marketing. We will focus on the ones that add value to staying on plan. The difference is that the marketing and sales professionals in the operational environment use these tools to *refine* forecasts, *adjust* estimates, and *analyze* data streams for adaptive purposes when goals are in jeopardy of not being met.

The "lean" way to gather data and make decisions in the post-launch environment is to use leading indicators of performance against a plan. You must measure lagging indicators in many cases, but let's be honest: They only let you know when you have to react to a problem that they have detected. Fire detectors work. The house burns while the fire department gets to the scene to contain the damage. Smoke detectors are better. They may warn you early enough to allow the smoldering source of the fire to be extinguished *before* significant damage occurs. What works best is a planned day-to-day way of behaving to prevent the fuel, air, and ignition sources from getting into the right "dynamic" combinations and starting a fire in the first place. Lean marketing and sales teams focus on their value-adding forms of fuel, air, and ignition sources so that the fundamentals, the dynamic elements, are measured and under adaptive control, preventing smoke and fire as much as possible.

Lean and Six Sigma rigor and discipline can help define these fundamentals. The Six Sigma for Growth approach is focused on identifying and documenting the integrated system of critical parameters that allow marketing and sales teams to measure and control the fundamentals.

Early warning data types and structures must be invented and used to enable a proactive approach to preventing sales and marketing problems. This requires innovation of measurement systems and data structures by marketing and sales teams. The best teams measure things their competitors do not even know exist as measurable forms of data. Some competitors may be aware of these metrics but find them difficult to measure. You must face this struggle and prevail. If you don't, you won't be able to sustain growth—just like everyone else.

Attaining balance between leading and lagging indicators of sales and marketing performance is a major goal in applying Six Sigma for Growth methods in post-launch line management. The key to successfully getting out ahead of marketing and sales problems is by measuring impending marketing and sales process failures. How can you see the impending failures? By developing innovative ways to measure things others are unwilling or unable to do. By gathering dynamic data sets and analyzing them as time-based samples to illuminate key trends in market and segment fundamentals that allow us to develop proprietary insights that our competitors lack.

As an example, let's explore applying Six Sigma for Growth to the classic product marketing market share development model of awareness-consideration-conversion (or hit rate) to measure the impact of post-launch sales promotion efforts. The classic awareness-consideration-conversion model seeks to develop a systematic way to promote awareness, consideration, and conversion of sales opportunities into revenue. Conversion is measured in terms of sales results after the fact. Applying Six Sigma for Growth adds key in-process measures to determine where to apply the elbow grease, such as establishing a goal of 30% awareness within four weeks of launch. Ultimately, you want to know what percentage of the target audience is aware of your new offering and tabulate the top ways in which potential customers became aware.

You can measure awareness quickly, inexpensively, and simply by sampling 20 to 50 people who represent your target audience, depending on the circumstances and population size. In this way, you can measure the overall impact of your market communications plan, understand the most important drivers of awareness, and tailor current and future communications plans for greater effectiveness and efficiency. Otherwise, overspending on marketing and sales tactics becomes a risk.

You might consider tracking the number of sales proposals currently in front of customers. This can help you understand if your sales funnel is adequately primed to deliver sales results at the current rate of conversion. It also can help you gauge how many proposals convert to sale, how many do not, and the underlying reasons why. This can provide solid feedback to the marketing process of what the customers value and what they don't. If you sell via channels, you might consider offering them online proposal-writing capabilities. Your channel partners may value this opportunity to differentiate their sales approach from competitors enough to share their sales funnel data with your company.

Among the biggest consumers of time and resources are trade shows because many companies "show up" to fulfill obligations rather than taking full advantage of the opportunities they present. If you have so many commitments that you just show up at these events, you may be doing more harm than good. Trade-show participation benefits from preparation that includes formal goal setting, marketing and sales tactics development, and metric development of both in-process (lead generation) and post-process (number of leads contacted and conversion rate) measurements to evaluate the trade show's return on investment. If you want to measure the impact of using a trade show to promote awareness, you might want to survey qualified targets as they leave the show floor. You may have sponsored a platform speaker at the conference, or distributed product samples or memorabilia, or bought advertising space on the back cover of the trade-show program. Which tactic best met your goal of creating awareness that ultimately elicited purchasing behavior? In this case, a multi-vari analysis could help answer that question. If the trade show novelty items included a toll-free phone number, the resulting call volume and conversions to sales could be used to determine the effectiveness of the item's ability to trigger a call to action.

The application of Six Sigma tools to the awareness-consideration-conversion model helps drive growth. The cause-and-effect analysis captures initial assumptions about what drives awareness. A customer behavior analysis maps how the target audience learns about offerings. Next, a model (such as $Y = f(x)$) is developed to define the cause-and-effect relationships of awareness activities on sales, for example. The KJ Analysis helps organize and communicate a large volume of potential customer "verbatim" feedback on the offering's benefits and positioning. Using the tools-tasks-deliverables structure, this chapter suggests how best to integrate Six Sigma for Growth tools into your current operational outbound marketing model.

Hard Versus Easy Data Sets

Effective control of the LMAD phases depends on what you choose to measure and how often. Many business processes have been designed and fitted with measurement systems that are loaded with things that are easy and convenient to measure. Often, easy-to-obtain data is least effective in telling you anything fundamental about what is really going on in your marketing and sales environments. Easy-to-measure data is almost always a lagging indicator of what has just happened. Just how lagging a measure is, is a function of how often you choose to take data and when you get around to analyzing it. It is common to hear teams state that they have tons of data but lack the time and resources to analyze it. This is where "lean" methods can help. Marketing and sales-Critical Parameter Management takes the hard fork in the road when it comes to the type of data, as well as the number and frequency of samples. To make a statistically sound decision, ensure that you collect the right data and collect enough of it. If you measure attribute data, you have to gather large amounts of data to make a sound statistical decision. If you measure continuous data, typically associated with a fundamental dynamic, relatively small samples of data are needed to make a statistically sound decision.

A good example of a poor measure of sales-force performance influence on revenue is the number of sales professionals sent out to interact with customers. Sheer numbers of people is a gross measure that is easy to quantify, but it can't provide behavioral relationship data in the context of human interaction. Qualitative relationship dynamics data is just plain hard to gather, document, and communicate, but it is one parameter that is fundamental to sales. If you can get the data, it will show a fundamental cause-and-effect relationship.

To stimulate innovation around data integrity and utility, you need to ask a few simple questions:

- Are you measuring continuous variables in your marketing and sales processes that are fundamental to the dynamics taking place within these processes?
- Are you measuring variables that your competitors are unaware of or are ignoring due to the difficulty of gathering such data?

Data is the result of taking measures while doing the work of marketing and sales. Data is produced and gathered by conducting tasks using specific tools, methods, and best practices.

The Tools, Methods, and Best Practices that Enable the LMAD Tasks

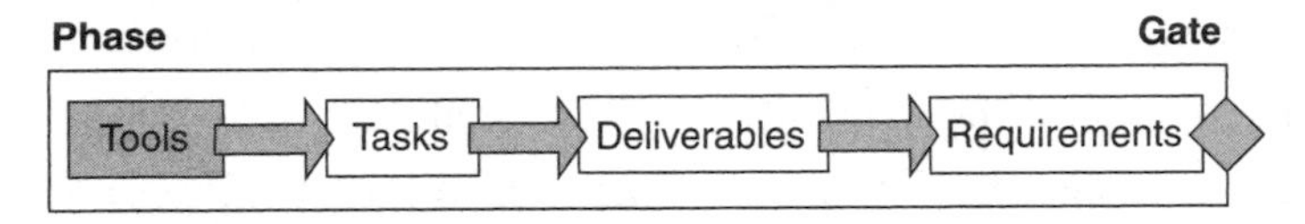

A *common set* of tools, methods, and best practices is used repeatedly during the LMAD phases, as described next. They enable the tasks of marketing and sales teams.

Process definition involves five major types of tools, methods, and best practices that are used to define the outbound marketing and sales process:

- Process requirements development includes interviewing and requirements data gathering; requirements structuring, ranking, and prioritization (such as KJ Analysis); and quality function deployment (to define detailed metrics).

- Process mapping captures the marketing and sales functions, inputs, outputs, and constraints in a process map, along with process noise mapping and Failure Modes & Effects Analysis (FMEA).
- Concept generation for marketing and sales professionals
- The Pugh Concept Evaluation & Selection Process, and the concept innovation process
- Critical Parameter Management entails critical parameter identification and metrics for product promotion; advertising; channel and distribution; customer relationship and support; and marketing communications.

Process risk management includes three major types of tools, methods, and best practices used to define the outbound marketing and sales process:

- Product/services management scorecard design and applications to drive risk analysis, risk management, and decision-making
- FMEA to delve further into risk analysis, risk management, and decision-making
- Strength Weakness Opportunity Threat (SWOT) Analysis to evaluate strengths, weaknesses, opportunities, and threats

Operational models, data analysis, and controls have several appropriate Six Sigma for Growth tools, methods, and best practices available:

- Line management control planning for the offering
- Project management methods, which borrow two key Six Sigma tools—cycle-time Monte Carlo Simulation and critical path task FMEA
- Marketing and sales process cost modeling

- Product and services price modeling
- Sales models and forecasting
- Market Perceived Quality Profiles (MPQP)
- Customer value chain mapping
- Design of surveys and questionnaires
- Post-launch data structures, analysis, and management utilizing four types of tools:
    1. Descriptive and inferential statistics such as graphical data mining, multi-vari studies, hypothesis testing, confidence intervals, t-Tests, data sample sizing, and last regression and model building
    2. Capability studies
    3. Statistical Process Control (SPC)
    4. Design of Experiments on the critical marketing and sales parameters, and conjoint analysis
- Adaptive price and sales forecasting methods using Monte Carlo Simulation
- Data feedback structures for advanced product/services planning
- Offering discontinuance planning

With the tools, methods, and best practices categorically defined, let's look at the phases of the LMAD process and explore how to use it to conduct marketing and sales tasks. In a post-launch steady-state environment, your business fluidly cycles in and out of the Manage and Adapt phases as necessary. Hence, a single scorecard format featuring the LMAD common tool set, methods, and best practices for your business can be adapted on an as-needed basis. Table 6.1 is a sample tools scorecard.

The sample scorecard, as shown in Table 6.1, is used by marketing teams that apply tools to complete tasks.

TABLE 6.1 Sample LMAD Tools Scorecard

1	2	3	4	5	6	7
SSFM Tool	**Quality of Tool Usage**	**Data Integrity**	**Results Versus Requirements**	**Average Tool Score**	**Data Summary, Including Type and Units**	**Task**
Process mapping flowcharts						Generate statistical process control charts/ capability indices
SPC/critical parameter management/capability studies						
Value chain mapping SIPOC/process requirements development methods/DOEs						Characterize critical processes and critical parameters in the launch environment
FMEA SWOT						Generate risk assessment
Noise diagrams						
R-W-W scorecards						
Sales modeling and forecasting methods/MPQ profiling/price modeling						Refine sales forecast
Capabilities studies regressions, ANOVA						Assess business case against current performance
Project management methods, FMEA, Monte Carlo simulation						Update control plan and FMEA data

Columns 1 and 7 align the tool to a specific task and its requirement for justifying its use. (Never use a tool if you can't justify its ability to fulfill a task.)

Quality of Tool Usage, Data Integrity, and Results Versus Requirements are scored on a scale of 1 to 10 using the simple scoring templates provided in Chapter 2, "Measuring Marketing Performance and Risk Accrual Using Scorecards."

Average Tool Score (ATS) illustrates overall tool quantification across the three scoring dimensions. (A high ATS indicates that a task is being underwritten by proper tool usage.)

Data Summary illustrates the nature of the data and the specific units of measure (such as attribute or continuous data, scale, or specific units of measure).

As we mentioned in Chapter 4, "Six Sigma in the Strategic Marketing Process," the following tools sections simply align the tools within each phase of the unique LMAD outbound marketing method to operationally manage a launched offering. The suggested tools draw from the complete set of Six Sigma/Lean tools readily available today; no "new" tools are introduced. Because this book is an executive overview, individual tool descriptions and guidelines on how to use them fall outside the scope of this book.

The Launch Phase

The marketing and sales teams have generated an integrated plan to see the launch through its cycle. The Launch phase is concluded when the product and/or services and support initiatives are fully distributed, available, and in a steady state mode for the Manage phase. The launch plan has a well-defined set of measurable control variables, functions, and results that are fundamental to the behavioral dynamics that characterize the unique issues associated with launching a new offering. The consistent application of a designed launch control plan has built-in robustness features to prevent excessive sensitivity to assignable, nonrandom sources of variation in the launch environment.

Requirements for the Launch Phase Milestone

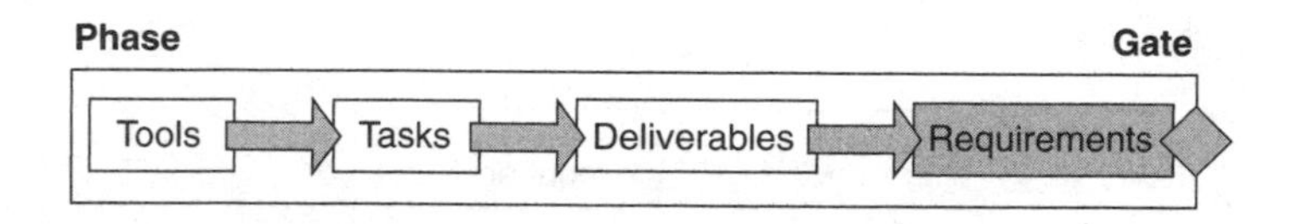

The milestone requirements for the launch involve the operational deployment and stabilizing the following two areas: product, services, and support resources; and marketing and sales processes, their detailed functions, and measurement systems (see Figure 6.4).

Tools

- Process definition
  - Requirements development
  - Process mapping
  - Concept generation
  - Pugh concept evaluation
  - Critical parameter management
- Process risk management
  - Score card
  - FMEA
  - SWOT
- Operational models, data analysis, and controls
  - Control planning
  - Project management
  - Cost modeling
  - Price modeling
  - Forecasting
  - MPQP
  - Value chain mapping
  - Surveys and questionnaires
  - Post-launch data analysis product discontinuance planning

Tasks

- Gather critical data
- Analyze critical performance data
- Generate statistical process control charts and capability indices
- Refine sales forecast models
- Generate LAUNCH phase risk assessment
- Assess business case against current performance
- Update control plan for the MANAGE phase

Deliverables

- Sales capability studies
- Customer identification and qualification metrics
- Customer purchase experience and behavioral analysis
- Customer satisfaction assessment
- Sales process failure modes and effects analysis
- Competitive assessment
- Advertising and promotion effectiveness evaluation
- Critical parameters update, including the SPC charts
- Business case fulfillment assessment
- Documented MANAGE phase control plan

FIGURE 6.4 The tools-tasks-deliverables for the launch phase.

Launch Phase Tasks and Deliverables

To establish operationally ready resources and processes, the following set of activities and their resulting deliverables must be delivered to fully satisfy the preceding Launch requirements.

Deliverables for the Launch Milestone

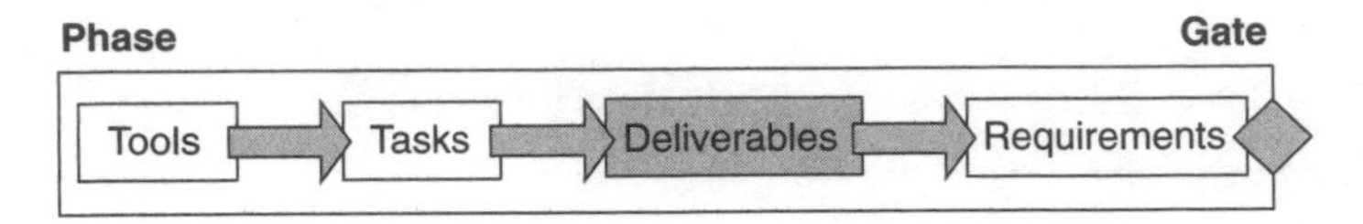

1. Sales capability studies to examine the actual sales data versus sales forecast, the initial sales trend assessment, and the sales growth rate versus the plan
2. Customer identification and qualification metrics
3. Customer purchase experience and behavioral analysis
4. Customer satisfaction assessment
5. Sales process FMEA
6. Competitive assessment
7. Advertising and promotion effectiveness evaluation
8. Critical parameters update, including the SPC charts
9. Business case fulfillment assessment
10. The documented Manage phase control plan is approved and ready for use

The sample scorecard, as shown in Table 6.2, is used by marketing gatekeepers who manage risk and make functional gate decisions for a specific project as part of the portfolio of projects being conducted by the marketing organization.

Columns 1 and 6 align the gate deliverable to a gate requirement. Each deliverable is justified as it contributes to meeting a gate requirement. (Never produce a deliverable if you can't justify its ability to fulfill a gate requirement.)

TABLE 6.2 Sample Launch Phase Deliverables Scorecard

1	2	3	4	5	6
SSFM Gate Deliverable	**Grand Average Tool Score**	**% Task Completion**	**Task Results Versus Deliverable Requirements**	**Risk Color (R, Y, G)**	**Key Milestone Requirements**
Marketing/sales capability study					Show category revenue growth versus industry/ actual sales versus forecast or initial plan/Cp and Cpk process metrics
Document customer identification/ qualification metrics					Produce a current market opportunity assessment that defines the market, segments, and targets
Customer purchase experience/behavioral analysis					Develop and show process maps, channel analysis, category online shopping, and statistical analysis of emerging trends
Customer satisfaction assessment					Develop a customer-perceived value analysis for each segment and specific targets within each segment
Sales and marketing process FMEA					Show risk assessments for sales, marketing, technical/ service, credit, distribution, communications, and so on
Market perceived quality gap analysis, Won-Lost Analysis, benchmark summary					Show a competitive assessment
Advertising and promotional effectiveness evaluation					Show how awareness, consideration, trial, usage, and referral will be promoted and measured
Critical parameters update					Show how CPM is fulfilling the control plan across critical sales and marketing processes
Business case fulfillment assessment					Show the results of the plan against actual results (for line extensions)
Documented control plan and FMEA data					Construct a control plan for post-launch marketing and sales processes with risk assessment

Grand Average Tool Score (GATS) illustrates aggregated tool quantification across the three scoring dimensions. (A high GATS indicates that a group of tasks is underwriting a gate deliverable.)

% Task Completion is scored on a 10 to 100% scale. This measure is critical if you want to understand how completely a group of related tasks are fulfilling a major gate deliverable.

Color coding illustrates the nature of the risk accrual for each major deliverable within this phase of the process. Color code risk definitions (red, yellow, and green) are found in Chapter 2.

Tasks Within the Launch Phase

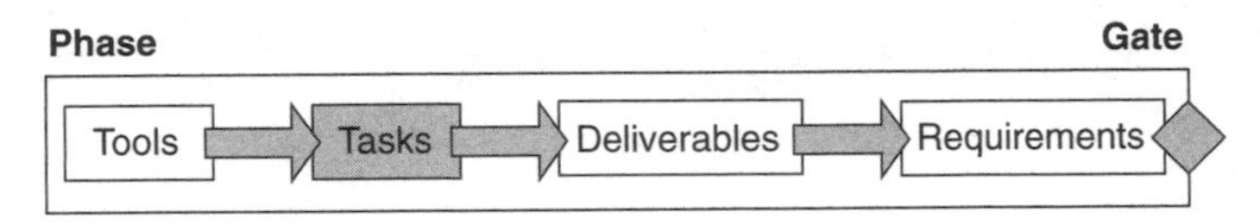

To produce the preceding set of deliverables, the following activities must be executed:

1. Generate Statistical Process Control charts and capability indices for key marketing and sales metrics using their data sets.
   - Gather critical marketing and sales data.
2. Characterize critical processes and parameters in launch environment.
   - Analyze critical marketing and sales performance data.
3. Generate a Launch phase risk assessment that includes conducting ongoing FMEA reviews within the Launch phase marketing and sales processes.
4. Refine sales forecast models.
5. Assess the business case against current performance.
6. Update the control plan for the Manage phase.

The sample scorecard, as shown in Table 6.3, is used by marketing project team leaders who manage major tasks and their timelines as part of their project management responsibilities.

TABLE 6.3 Sample Launch Phase Task Scorecard

1	2	3	4	5	6
SSFM Task	**Average Tool Score**	**% Task Completion**	**Task Results Versus Gate Requirements**	**Risk Color (R, Y, G)**	**Key Milestones**
Generate statistical process control charts/ capability indices					Marketing/sales capability study
					Customer purchase experience/ behavioral analysis
					Critical parameters update
Characterize critical processes and parameters in launch environment					Document customer identification/ qualification metrics
					Customer satisfaction assessment
Generate risk assessment (FMEA)					Risk analysis of value chain processes (including sales)
Refine sales forecast					Business case fulfillment assessment
Assess business case against current performance					Business case fulfillment assessment
					Advertising and promotions effectiveness evaluation
					Competitive assessment
Update control plan					Documented control plan

Columns 1 and 6 align the task to a specific gate deliverable for justifying the task. (Never conduct a task if you can't justify its ability to produce a gate deliverable and fulfill a gate requirement.)

Average Tool Score (ATS) illustrates overall tool quantification across the three scoring dimensions. (A high ATS indicates that a task is being underwritten by proper tool usage.)

% Task Completion is scored on a 10 to 100% scale. This measure is critical if you want to understand how completely each specific, major task is being done.

Color coding illustrates the nature of the risk accrual for each major task within this phase of the process. Color code risk definitions are found in Chapter 2.

The Manage Phase

Once the launch cycle has been completed, the steady-state control plan kicks into action to manage the new offering through its life cycle. A steady-state marketing and sales plan consists of key tasks that are enabled by specific tools, methods, and best practices to sustain the offering as it enters a mature, stable marketing and sales environment. Marketing and sales process critical parameters are measured and evaluated to determine when the Adapt phase is required to sustain sales goals. Depending on the offering's life cycle, both the Manage and Adapt phases can repeat several times before the final phase, Discontinue, is reached.

Requirements for the Manage Phase Milestone

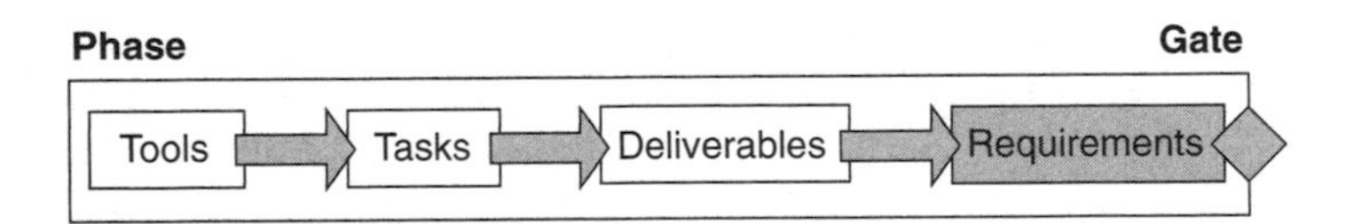

This milestone requirement is simply that the marketing and sales processes, with their detailed functions and measurement systems, *can detect assignable causes of variation* that indicate when adaptive actions are required to stay on plan (see Figure 6.5).

Tools

- Process definition
  - Requirements development
  - Process mapping
  - Concept generation
  - Pugh concept evaluation
  - Critical parameter management
- Process risk management
  - Score card
  - FMEA
  - SWOT
- Operational models, data analysis, and controls
  - Control planning
  - Project management
  - Cost modeling
  - Price modeling
  - Forecasting
  - MPQP
  - Value chain mapping
  - Surveys and questionnaires
  - Post-launch data analysis product discontinuance planning

Tasks

- Gather critical data
- Analyze critical performance data
- Generate Statistical Process control charts and capability indices
- Refine sales forecast models
- Generate LAUNCH phase risk assessment
- Assess business case against current performance
- Update control plan for the MANAGE phase

Deliverables

- Sales capability studies
- Customer identification and qualification metrics
- Customer purchase experience and behavioral analysis
- Customer satisfaction assessment
- Sales process failure modes and effects analysis
- Competitive assessment
- Advertising and promotion effectiveness evaluation
- Critical parameters update, including the SPC charts
- Business case fulfillment assessment
- Documented MANAGE phase control plan

FIGURE 6.5 The tools-tasks-deliverables for the Manage phase.

Manage Phase Tasks and Deliverables

To manage ongoing operational marketing and sales processes, *the same set of Launch tasks and resulting deliverables are used in the Manage phase*. The difference is that they continue to get updated and modified as the environment and information change over time.

The Adapt Phase

This phase addresses how to determine when to react or be proactive to events to stay on plan. As mentioned in Chapter 5, “Six Sigma in the Tactical Marketing Process,” a well-constructed plan identifies the critical parameters to monitor in a steady-state process. Such a plan also defines a set of contingencies, enabled by specific sets of tools, methods, and best practices, that determine when and how to respond to change or an off-plan trend. This contingencies plan identifies the countermeasures to assignable causes of variation that disrupt the steady-state plan for growth. Changes in economic, industry, or market environments may cause an adaptive response that should be unique to your business and your offering.

Changes in economic conditions provide examples of the importance of developing a potential adaptive response. For example, when fuel prices escalate, transportation companies should have a control plan in place to inform operations how best to respond. A one-cent fuel increase costs American Airlines $80 million annually and perhaps is worthy of triggering an adaptive strategy in ticket pricing, in-flight services, or another approach. Fluctuations in the currency exchange rate are another example of an economic condition affecting multinational companies. The Toronto Blue Jays baseball team is dependent on either traveling to the U.S. or hosting Americans. It probably has to plan a currency hedge on how currency fluctuations affect attendance and concessions. Also, when the U.S.-to-Canadian exchange rate is low, the Blue Jays may execute a response strategy to attract top athletes because paying salaries may be easier at that point. Such an adaptive response fits nicely with an overall strategy to build a team of top athletes, which presumably produces better team performance that yields higher ticket, parking, and food revenues.

Sources of *assignable cause variation* (nonrandom noises) signal the marketing and sales teams to adapt or make adjustments to critical adjustment parameters that are known to be capable of countering these effects. These variations typically are changes in customer behavioral dynamics or market dynamics that threaten your results. The goal is not to change your plan but to adaptively use it to stay in control of growth parameters. The Adapt phase can be exited by bringing the marketing and sales metrics back to a steady state of control after the effects of variation have been countered.

Requirements for the Adapt Phase Milestone

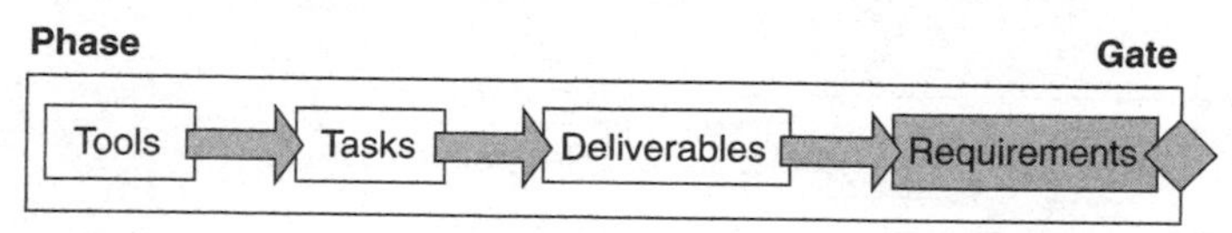

The Adapt phase contains two key requirements. The first is *identifying critical adjustment parameters* that can return the marketing and sales processes to a state of statistical control. Second, marketing and sales *identify the leading indicators (assignable causes of variation)* that signal the marketing and sales teams to adjust their critical adjustment parameters (see Figure 6.6).

Tools

- ◆ Process definition
  - Requirements development
  - Process mapping
  - Concept generation
  - Pugh concept evaluation
  - Critical parameter management
- ◆ Process risk management
  - Score card
  - FMEA
  - SWOT
- ◆ Operational models, data analysis, and controls
  - Control planning
  - Project management
  - Cost modeling
  - Price modeling
  - Forecasting
  - MPQP
  - Value chain mapping
  - Surveys and questionnaires
  - Post-launch data analysis product discontinuance planning

Tasks

- ◆ Apply critical adjustment parameters to adjust critical functions and results back on target
- ◆ Conduct designed experiments as necessary
- ◆ Generate Statistical Process Control charts and capability indices after adjustments
- ◆ Refine price models
- ◆ Refine advertising, promotion, and channel management plans
- ◆ Refine sales forecast models
- ◆ Generate ADAPT phase risk assessment
- ◆ Assess business case against current performance
- ◆ Refine ADAPT phase control plan
- ◆ Refine DISCONTINUE phase control plan

Deliverables

- ◆ Updated critical parameters
- ◆ Updated process noise maps
- ◆ Updated process FMEAs
- ◆ Documented DISCONTINUE phase plan

FIGURE 6.6 The tools-tasks-deliverables for the Adapt phase.

Adapt Phase Tasks and Deliverables

To establish operationally adaptive resources and processes, the following four activities and their resulting deliverables help meet the Adapt phase requirements.

Deliverables for the Adapt Milestone

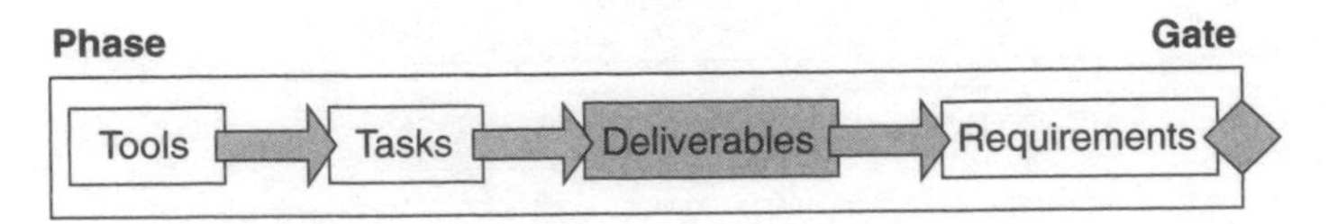

1. Updated critical parameters, including the SPC charts and critical adjustment parameters.
2. Updated process noise maps.
3. Updated process FMEAs.
4. Updated annual operating plan.
5. Documented Discontinue phase plan approved and ready for use.

The sample scorecard, as shown in Table 6.4, is used by marketing gatekeepers who manage risk and make functional gate decisions for a specific project as part of the portfolio of projects being conducted by the marketing organization.

Columns 1 and 6 align the gate deliverable to a gate requirement. Each deliverable is justified as it contributes to meeting a gate requirement. (Never produce a deliverable if you can't justify its ability to fulfill a gate requirement.)

Grand Average Tool Score (GATS) illustrates aggregated tool quantification across the three scoring dimensions. (A high GATS indicates that a group of tasks is underwriting a gate deliverable.)

% Task Completion is scored on a 10 to 100% scale. This measure is critical if you want to understand how completely a group of related tasks are fulfilling a major gate deliverable.

Color coding illustrates the nature of the risk accrual for each major deliverable within this phase of the process. Color code risk definitions are found in Chapter 2.

TABLE 6.4 **Sample Adapt Phase Deliverables Scorecard**

1	2	3	4	5	6
SSFM Gate Deliverable	**Grand Average Tool Score**	**% Task Completion**	**Task Results Versus Gate Deliverable Requirements**	**Risk Color (R, Y, G)**	**Key Milestones**
Updated critical parameters, SPC charts, and critical adjustment parameters					Identify critical adjustment parameters that can return marketing and sales processes to a state of statistical control
Updated process noise maps					Identify leading indicators (assignable causes of variation) showing when to adjust critical parameters
Updated process FMEAs					Identify which processes are most likely to fail and a plan of action that will be executed in the event of a failure
Updated annual operating plan					Show the market situation analysis; marketing plan; and schedule of actions, costs, and responsibilities
Documented Discontinue phase plan					Show the plan and process that will be used to discontinue the offering

Tasks Within the Adapt Phase

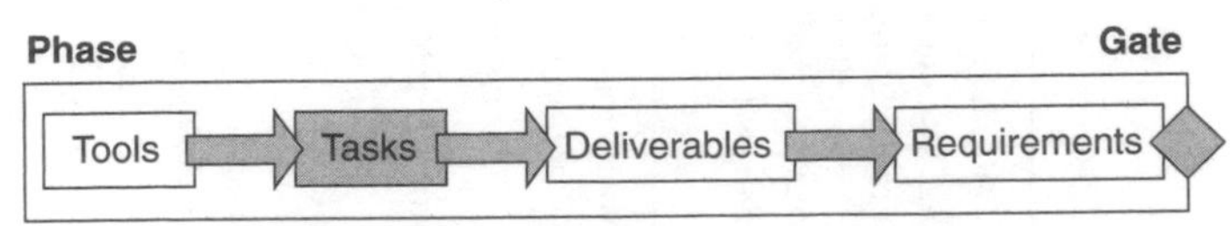

1. Update critical parameters, Statistical Process Control charts, and critical adjustment parameters.
   a. Apply critical adjustment parameters to adjust critical marketing and sales functions and get their results back on target.
   b. Conduct designed experiments as necessary to improve the effectiveness of critical adjustment parameters for current marketing and sales conditions.
2. Conduct DOEs and generate SPC control charts.
   a. Conduct designed experiments as necessary to improve effectiveness of critical adjustment parameters for current marketing and sales conditions.
   b. Generate Statistical Process Control charts and capability indices after adjustments.
3. Develop Noise Maps.
4. Update process FMEAs, by updating the competitive situation analysis, conducting a SWOT Analysis (using the results of the critical parameter adjustments), and conducting market perceived quality profile and gap analysis.
5. Refine price models.
6. Refine advertising, promotion, and channel management plans.
7. Refine sales forecast models.
8. Develop Discontinue phase plan and risk assessment by assessing the business case against current performance, and refining the Control plan.

The sample scorecard, as shown in Table 6.5, is used by marketing project team leaders who manage major tasks and their timelines as part of their project management responsibilities.

TABLE 6.5 Sample Adapt Phase Task Scorecard

1	2	3	4	5	6
SSFM Task	**Average Tool Score**	**% Task Completion**	**Task Results Versus Gate Requirements**	**Risk Color (R, Y, G)**	**Key Milestones**
Update critical parameters, SPC charts, and critical adjustment parameters					Updated critical parameters, SPC charts, and critical adjustment parameters
Conduct DOEs and generate SPC control charts					Identified new critical adjustment parameters, if any
Develop noise maps					Updated process noise maps
Update process FMEAs					Updated process FMEAs
Refine price models					Updated annual operating plan
Refine advertising, promotion, and channel management plans, as necessary					
Refine sales forecast models					
Develop Discontinue phase plan and risk assessment					Documented Discontinue phase

Columns 1 and 6 align the task to a specific gate deliverable for justifying the task. (Never conduct a task if you can't justify its ability to produce a gate deliverable and fulfill a gate requirement.)

Average Tool Score (ATS) illustrates overall tool quantification across the three scoring dimensions. (A high ATS indicates that a task is being underwritten by proper tool usage.)

% Task Completion is scored on a 10 to 100% scale. This measure is critical if you want to understand how completely each specific, major task is being done.

Color coding illustrates the nature of the risk accrual for each major task within this phase of the process. Color code risk definitions are found in Chapter 2.

The Discontinue Phase

As the product (or service) nears its planned or forced end of life, a preplanned set of deliverables and tasks will increase the likelihood of an efficient, more cost-effective transition to new products and services as you renew your offerings portfolio.

Requirements for the Discontinue Phase Milestone

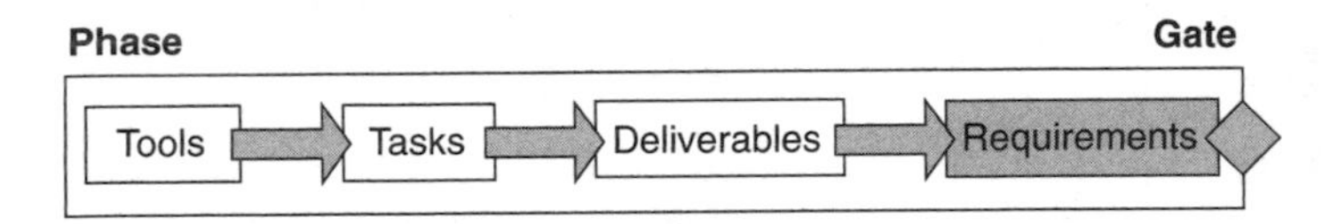

This final phase has three simple requirements to complete the closure (see Figure 6.7):

1. Define market and sales conditions that fit discontinuance criteria.
2. Discontinue products and/or services according to the discontinuance plan.
3. Provide discontinuance data to the portfolio renewal team.

Tools

- Process definition
  - Requirements development
  - Process mapping
  - Concept generation
  - Pugh concept evaluation
  - Critical parameter management
- Process risk management
  - Score card
  - FMEA
  - SWOT
- Operational models, data analysis, and controls
  - Control planning
  - Project management
  - Cost modeling
  - Price modeling
  - Forecasting
  - MPQP
  - Value chain mapping
  - Surveys and questionnaires
  - Post-launch and data analysis product discontinuance planning

Tasks

- Apply critical adjustment parameters to adjust critical functions and their results
- Conduct designed experiments as necessary
- Refine price models for discontinuance
- Refine advertising, promotion, and channel management plans for discontinuance
- Generate discontinuance sales forecast models
- Generate data back to the product portfolio renewal team
- Conduct final business case assessment against goals

Deliverables

- Critical parameter data
- Discontinuance phase risk assessment
- Assess business case against current performance
- Recommendations for sustaining the brand through next generation product and services

FIGURE 6.7 The tools-tasks-deliverables for the Discontinue phase.

Discontinue Phase Tasks and Deliverables

To transition resources and processes to an end-of-life state and prepare for the transition to a new offering, the following activities and their resulting deliverables help meet the Discontinue phase requirements.

Deliverables for the Discontinue Milestone

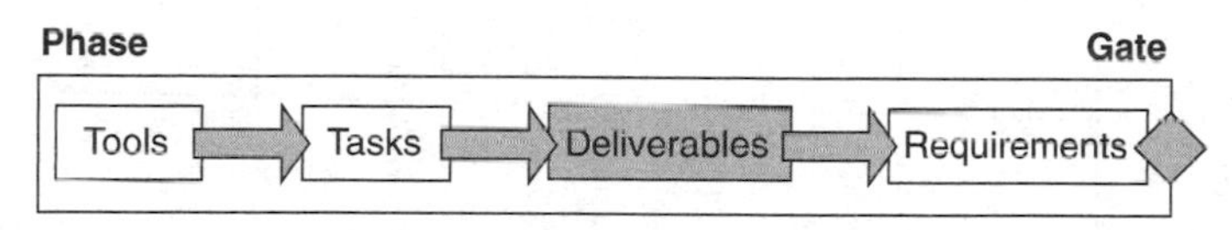

1. Documented discontinuance criteria using the critical parameter data (including SPC charts, capability studies, and trend analysis) to assess the business case against current performance.

TABLE 6.6 Sample Discontinue Phase Deliverables Scorecard

1	2	3	4	5	6
SSFM Gate Deliverable	**Grand Average Tool Score**	**% Task Completion**	**Task Results Versus Gate Deliverable Requirement**	**Risk Color (R, Y, G)**	**Key Milestones**
Documented discontinuance criteria					Define discontinuance criteria
Discontinue phase project plan and risk assessment					Show thorough plans with process maps for each critical process—communications, ongoing supply of parts and labor, warranty, recycling, and so on
					Must show FMEA
Documented discontinuance data sent to portfolio renewal team					Provide discontinuance data to portfolio renewal team

2. Discontinue phase project plan and risk assessment (via updating the process FMEA, noise maps, and competitive assessments).

3. Documented discontinuance data sent to portfolio renewal team. Include recommendations for sustaining the brand through next-generation products and services, combining data gathered from the voice of marketing, Voice of the Customer, voice of the sales, market perceived quality profile and gaps, the final SWOT, Porter's 5 Forces Analysis, and the lessons learned document.

The sample scorecard, as shown in Table 6.6, is used by marketing gatekeepers who manage risk and make functional gate decisions for a specific project as part of the portfolio of projects being conducted by the marketing organization.

Columns 1 and 6 align the gate deliverable to a gate requirement. Each deliverable is justified as it contributes to meeting a gate requirement. (Never produce a deliverable if you can't justify its ability to fulfill a gate requirement.)

Grand Average Tool Score (GATS) illustrates aggregated tool quantification across the three scoring dimensions. (A high GATS indicates that a group of tasks is underwriting a gate deliverable.)

% Task Completion is scored on a 10 to 100% scale. This measure is critical if you want to understand how completely a group of related tasks are fulfilling a major gate deliverable.

Color coding illustrates the nature of the risk accrual for each major deliverable within this phase of the process. Color code risk definitions are found in Chapter 2.

Tasks Within the Discontinue Phase

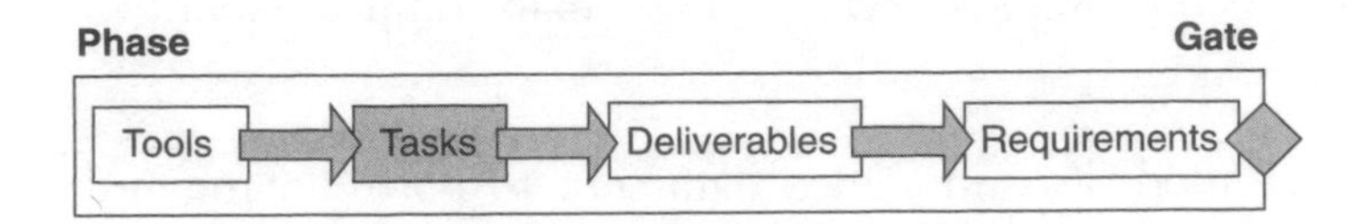

1. Apply critical adjustment parameters to adjust critical marketing and sales functions and their results to control the discontinuance of the products and services to protect the brand and maximize the business case. May conduct designed experiments as necessary to improve the effectiveness of critical adjustment parameters for discontinuance conditions.
2. Refine price models for discontinuance.
3. Refine advertising, promotion, and channel management plans for discontinuance.
4. Generate discontinuance sales forecast models.
5. Generate and provide marketing and sales data to the portfolio renewal team by conducting SWOT and Porter's 5 Forces Analysis, MPQP and gap analyses, competitive assessments, and voice of "X" data gathering. Also generate a lessons learned document to give to the portfolio renewal team.
6. Conduct a final business case assessment against goals.

The sample scorecard, as shown in Table 6.7, is used by marketing project team leaders who manage major tasks and their timelines as part of their project management responsibilities.

Columns 1 and 6 align the task to a specific gate deliverable for justifying the task. (Never conduct a task if you can't justify its ability to produce a gate deliverable and fulfill a gate requirement.)

Average Tool Score (ATS) illustrates overall tool quantification across the three scoring dimensions. (A high ATS indicates that a task is being underwritten by proper tool usage.)

TABLE 6.7 Sample Discontinue Phase Task Scorecard

1	2	3	4	5	6
SSFM Task	**Average Tool Score**	**% Task Completion**	**Task Results Versus Gate Requirements**	**Risk Color (R, Y, G)**	**Gate Deliverable**
Apply critical adjustment parameters to control the discontinuance, protect the brand, and maximize the business case					Discontinue phase project plan and risk assessment
Refine price models for discontinuance					
Refine advertising, promotion, and channel management plans for discontinuance					
Generate discontinuance sales forecast models					
Generate and provide marketing and sales data to portfolio renewal team					
Conduct final business case adjustment against goals					

% Task Completion is scored on a 10 to 100% scale. This measure is critical if you want to understand how completely each specific, major task is being done.

Color coding illustrates the nature of the risk accrual for each major task within this phase of the process. Color code risk definitions are found in Chapter 2.

Summary

The LMAD operational process is now complete. We have illustrated how Six Sigma can be used to help control, adapt, and refine what you do and when you do it in these phases. It is worth noting that the Manage and Adapt phases can repeat numerous times before Discontinue is conducted. For example, the jeans industry experienced a steady decline in sales until the launch of Calvin Klein designer jeans, which reinvigorated the industry category from mature into the early stages of a sustained, profitable growth curve. How long a product and its services can generate revenue and margin in fulfillment of the business case is a variable that can be forecast and assessed with the help of the Six Sigma Tools, methods, and best practices.

7

Quick Review of Traditional DMAIC

The Classic Six Sigma Approach of Problem-Solving in a Steady-State Process or an Existing Product and Service

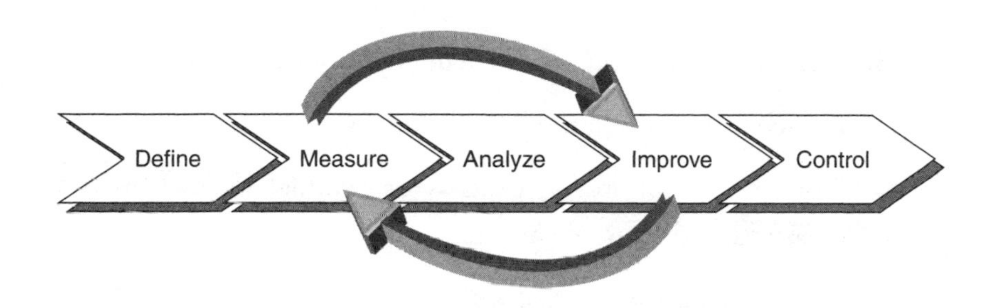

The Traditional Six Sigma DMAIC Method

Several books have been written on the traditional Six Sigma DMAIC method. However, to provide you with a complete overview of Six Sigma concepts that are suitable for marketing, this chapter provides a quick review of the DMAIC (pronounced "duh-may-ick") method's tools, tasks, and deliverables. This approach is often called the *process improvement* methodology. It was designed to solve a problematic process or product and/or service offering to regain control of its cost, quality, and/or cycle time. Fixes often include defects or failures, excess steps, and deterioration. This method was designed to detect a problem with an *existing steady-state process or offering*.

The DMAIC methodology uses a process-step structure. Steps generally are sequential; however, some activities from various steps may occur concurrently or may be iterative. Deliverables for a given step must be completed prior to formal gate review approval. Step reviews occur sequentially. The five steps of DMAIC are as follows:

1. **Define** the problem.
2. **Measure** the current process or product performance.
3. **Analyze** the current performance to isolate the problem.
4. **Improve** the problem by selecting a solution.
5. **Control** the improved process or product performance to ensure that the targets are met.

A DMAIC project typically is relatively short in duration (three to nine months) versus product development projects (using UAPL or DFSS) and operational line management (using LMAD), which can run years. Given the relatively shorter duration of other types of Six Sigma methodologies, we distinguish DMAIC as having five *steps* rather than phases. Figure 7.1 illustrates the five-step model and its iterative nature. It is used throughout this chapter to help identify the particular step being highlighted.

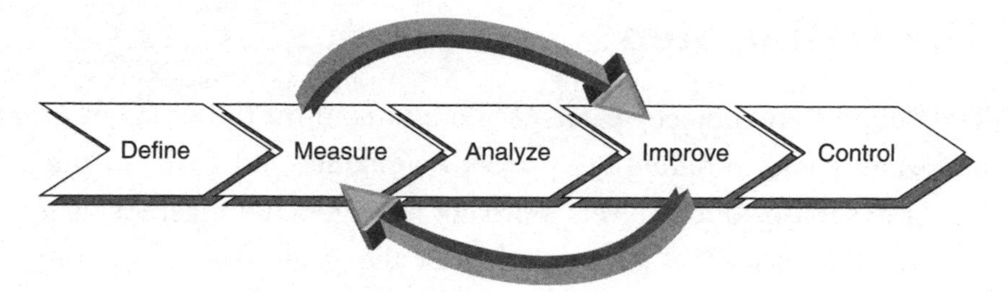

FIGURE 7.1 DMAIC process steps.

The DMAIC method is primarily based on the application of *statistical process control*, *quality tools*, and *Process Capability Analysis*; it is *not* a product development methodology. It can be used to help redesign a process—any process—given that the redesign fixes the initial process problem. To be implemented, the method requires four components:

- A measurement system (a gauge)
- Statistical analysis tools (to assess samples of data)
- An ability to define an adjustment factor (to put the response on target)
- A control scheme (to audit the response performance against statistical control limits)

Given that these four components are in place, let's quickly review each of the five steps, their requirements, and typical tool-task-deliverable combinations.

As mentioned in Chapter 4, "Six Sigma in the Strategic Marketing Process," this book provides an executive overview of a methodology by simply aligning the appropriate tools to a given phase—or, in this case, methodology step. Hence, individual tool descriptions and guidelines on how to use them fall outside the scope of this book. In addition, it is important to mention that any given Six Sigma project may or *may not* use all the tools aligned within a given phase or step; it depends on the context and the project's complexity. Hence, it is important to understand how and when to use the tools to ensure that *the right tool is used at the right time to answer the right question.*

The Define Step

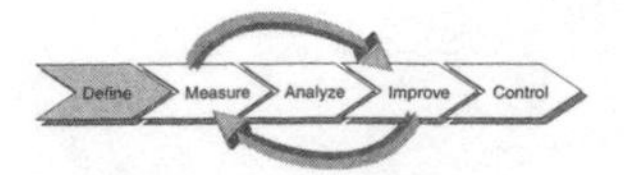

The define step's objectives are twofold: to confirm the problem or opportunity and to define the project boundaries and goals. In this step, the sponsor often selects a project manager, and together they confirm the project objectives, identify the goals (in measurable terms), and create a high-level project plan to kick off the project.

Define Step Tools, Tasks, and Deliverables

The Define step is critical to anchor the project. Clear identification of what is in scope and out of scope are critical deliverables. This step involves identifying the project stakeholders and project team members; hence, the ever-critical communication begins in this first step.

Gathering the facts about the problem (or opportunity) and taking the time to plan are important to establish the correct course for the project. Hence, the Define activities focus on defining the project scope. Skimping on these tasks or deliverables could lead to an out of control project and misaligned fix to the problem. The tools allow the tasks to produce the right deliverables to meet the Gate requirements and ensure measurable results. Figure 7.2 summarizes the tools, tasks, and deliverables for the Define step.

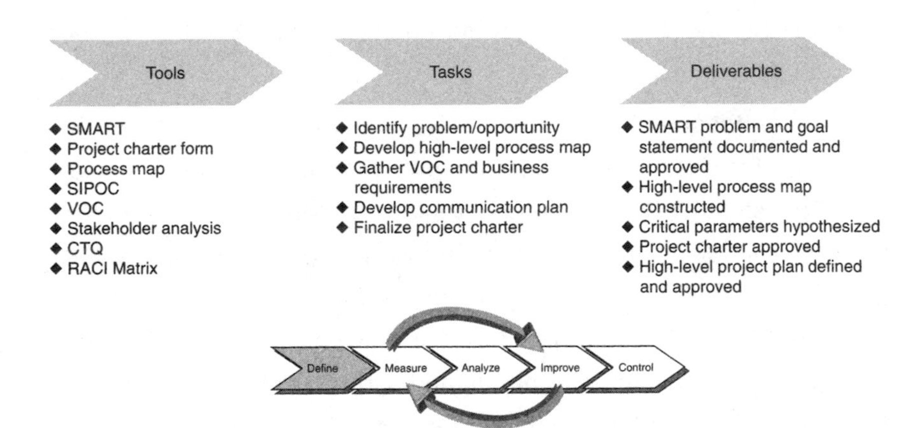

FIGURE 7.2 The tool-task-deliverables combination for the Define step.

Define Step Deliverables

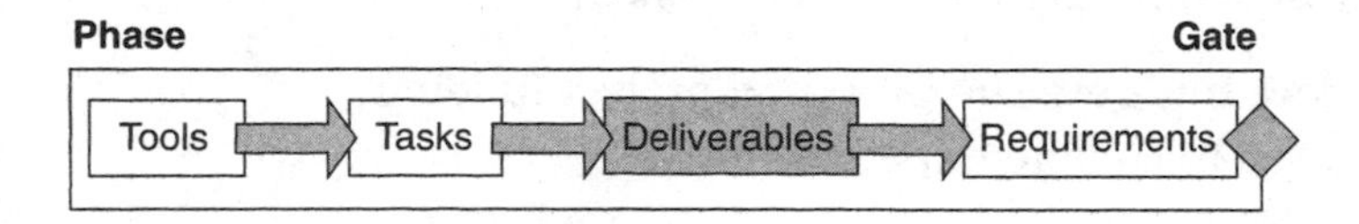

The five deliverables in the Define step are as follows:

1. **SMART problem and goal statement.** The problem statement describes *what is wrong* or the *opportunity for improvement*, and the goal statement defines the team's improvement objective. The problem statement highlights the current situation and avoids proposing a solution or answer for how to achieve the desired state. Both need to be SMART—Specific, Measurable (from the present level to the target level), Achievable (yet aggressive), Relevant to the project team and business, and Time-bounded. Using the SMART approach helps focus the team effort and verifies that the team members and managers are aligned on the critical improvement efforts.

2. **High-level process map,** depicting about three to five summary process steps and establishing the project boundaries.

3. **Critical parameters hypothesized** as to what may be causing the undesired results.

4. **Approved project charter** summarizing the project and defining the following elements:

   Overview: Alignment with overall company strategy/vision.

   Opportunity statement: Problem or opportunity statement and VOC target.

   Goal statement: Approach on how to achieve or solve the opportunity statement.

   Deliverables: Define the project outputs.

Audiences covered: What's in scope.

Out of scope: Define project boundaries.

Project budget: The proposed project funding required to complete the deliverables.

Key stakeholders: Executive sponsor, project team, subject matter experts.

Project time line: Proposed project milestones (including milestone review meetings) and a target completion date for each.

5. **Approved high-level project plan** defining a summary project time line.

Define Step Tasks

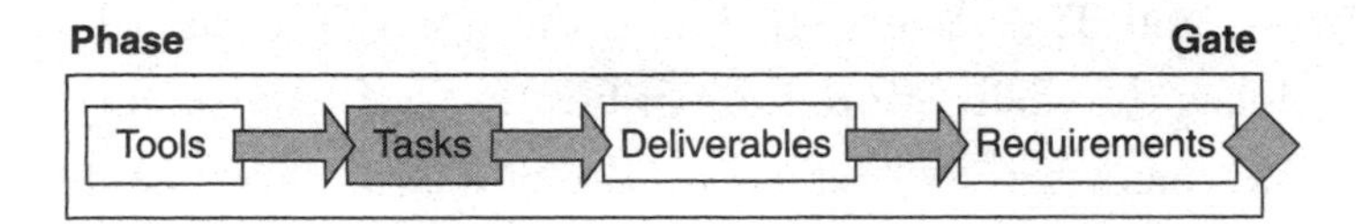

The Define tasks closely parallel the step's eight outputs:

1. Identify and scope (quantify) the problem or opportunity.
2. Develop a high-level process map.
3. Gather VOC and business requirements.
4. Develop a communication plan.
5. Finalize the project charter.
6. Select a project sponsor and members.
7. Identify stakeholders.
8. Gain approval and necessary funding to conduct the DMAIC project.

Tools That Enable the Define Tasks

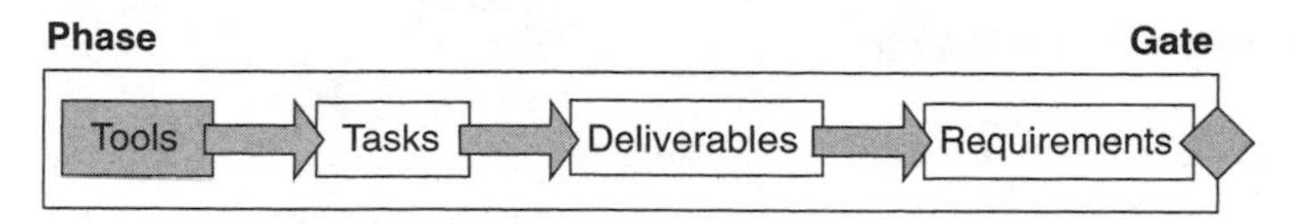

The tool set that supports the Define tasks to provide the Define deliverables includes the following software, methods, and best practices:

- **SMART** method to define the problem statement and project goals within the context of Specific, Measurable, Achievable, Relevant, and Time-bounded
- **Project charter form**
- **Process map**
- **SIPOC:** A summary tool to communicate the process's Suppliers, Inputs, Process, Outputs, and Customers
- **VOC:** Voice of the Customer
- **Stakeholder analysis**
- **CTQ:** Critical To Quality, according to the customers
- **RACI Matrix:** Responsible, Accountable, Consulted, Informed

The Measure Step

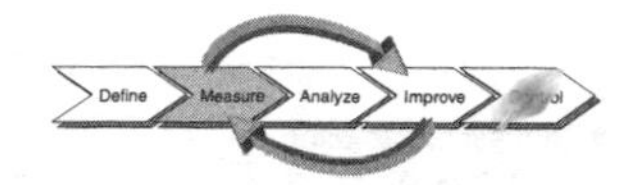

The Measure step gathers the necessary data to understand and measure the current state. This step identifies the current state's magnitude and units and evaluates whether the existing data collection system is reproducible and repeatable. Collecting or establishing the baseline of current state performance and process metrics is the crux of the Measure step. Depending on the project scope (size and complexity) and the rigor of the current state monitoring, this step can be long and involved if critical data is missing.

The second step of DMAIC focuses on three primary objectives:

1. Selecting CTQ characteristics.
2. Defining the performance standards.
3. Validating the current measurement system.

At the Measure step review, in addition to inspecting whether the objectives have been met, a detailed process map often is requested. The next level of information on a process map provides a means to begin identifying potential non-value-adding activities as candidates for immediate action. As part of this review and subsequent reviews, the project sponsor and manager also should verify the project performance—is it on plan, do any barriers (or potential barriers) exist, and is "scope creep" being held at bay?

Measure Step Tools, Tasks, and Deliverables

The tools enable the tasks to produce the deliverables that meet the Measure step requirements and ensure measurable results. Figure 7.3 summarizes the tools, tasks, and deliverables for the Measure step.

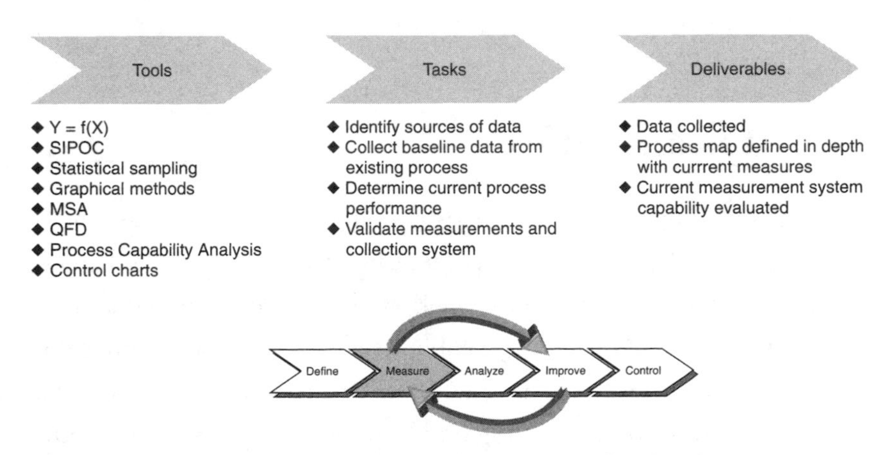

FIGURE 7.3 The tool-task-deliverables combination for the Measure step.

Measure Step Deliverables

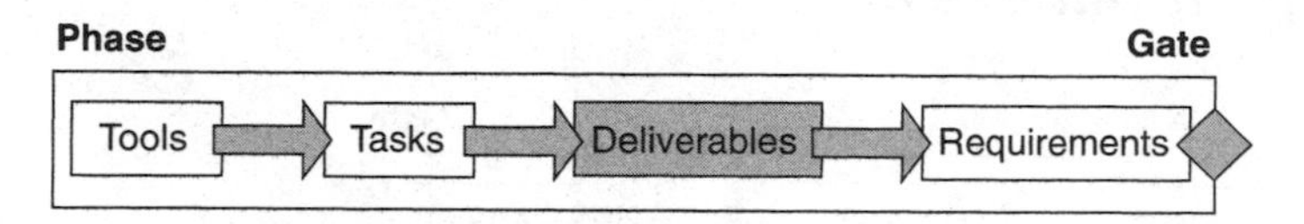

The three major deliverables for the Measure step are as follows:

1. **Data collected** on the current process and performance metrics.
2. **Detailed process map** defined in depth with current measures.
3. **Current measurement system capability evaluated.**

Measure Step Tasks

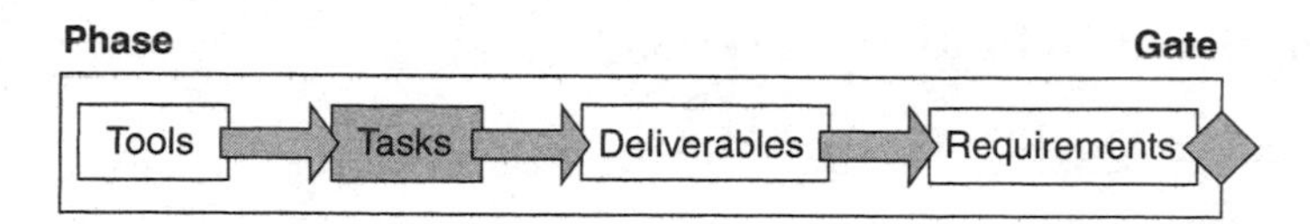

The Measure tasks that produce this step's outputs are as follows:

1. Identify sources of data.
2. Collect baseline data from the existing process.
3. Determine the current process's performance.
4. Validate measurements and the collection system:
   a. Gather the data collection plan and the current state process and performance data.
   b. Identify potential process variations.

Tools, Methods, and Best Practices That Enable the Measure Tasks

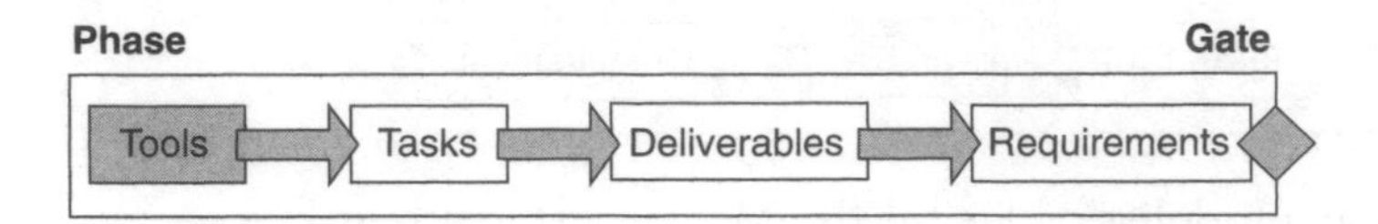

The tool set that supports the Measure tasks to provide the Measure deliverables includes the following software, methods, and best practices:

- **Y = f(X)**
- **Suppliers, Inputs, Process, Outputs, and Customers**—updated or revised as appropriate
- **Detailed process map**
- **Operational Definition Worksheet**
- **Critical to Quality Tree or Matrix**
- **Data Collection Worksheet**
- **Statistical sampling**
- **Graphical methods**
- **Measurement System Analysis**
- **Quality Function Deployment**
- **Process Capability Analysis**
- **Control charts**

The Analyze Step

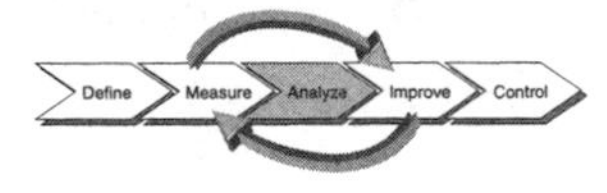

The Analyze step's objective is to find the *root cause* of the problem. In this step, the project team delves into the details to enhance its understanding of the process (or product/service offering) and the problem. They draw on analytical tools to dissect the *root cause* of process variability and separate the vital few inputs from the trivial many. The team's activity level increases during this step.

The Analyze step requirements are as follows:

- Interpret the data to establish any cause-and-effect relationships:
  - Analyze the data for patterns.
  - Prioritize by magnitude of impact.
  - Identify gaps.
- Select potential improvement and innovation opportunities to explore further.

Analyze Step Tools, Tasks, and Deliverables

The tools enable the tasks to produce the deliverables that meet the Analyze step requirements and ensure measurable results. Figure 7.4 summarizes the set of tools, tasks, and deliverables for the Analyze step.

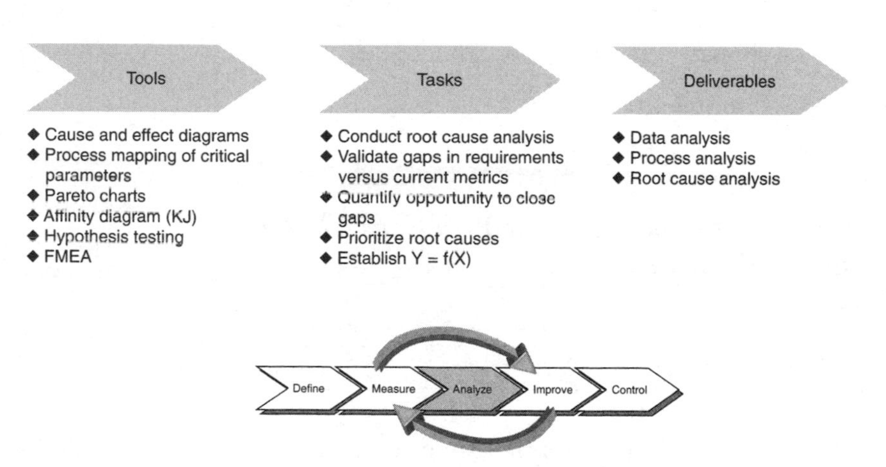

FIGURE 7.4 The tool-task-deliverables combination for the Analyze step.

Analyze Step Deliverables

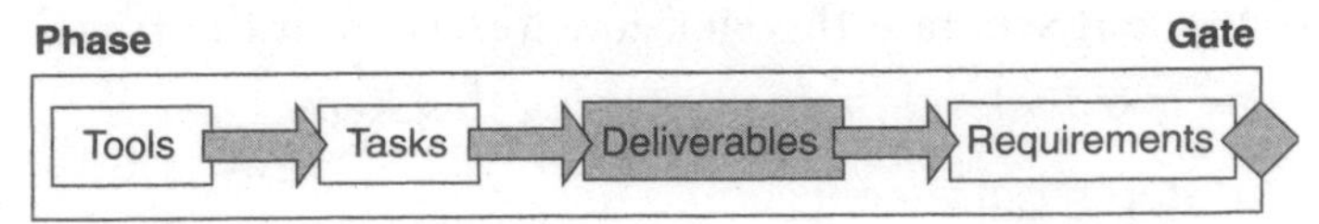

The three Analyze step deliverables are equally important; no one output should be overlooked or incomplete. They include the following:

1. **Data analysis**, which identifies the vital few.
2. **Process analysis**, which identifies the non-value-adding steps and process failures in the process.
3. **Root cause analysis** to establish the potential cause-and-effect relationship.

Analyze Step Tasks

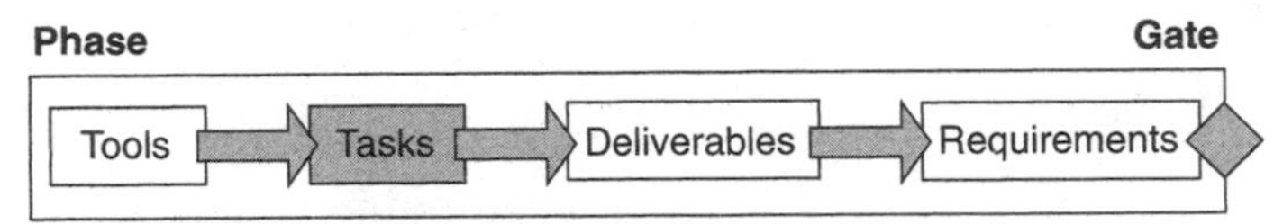

Dictated by the Analyze deliverables required to complete the third DMAIC step, the tasks within the Analyze step include the following:

1. Conduct a root cause analysis.
2. Validate gaps in requirements versus current metrics.
3. Quantify an opportunity to close gaps.
4. Prioritize root causes.
5. Establish $Y = f(X)$.

Tools, Methods, and Best Practices That Enable the Analyze Tasks

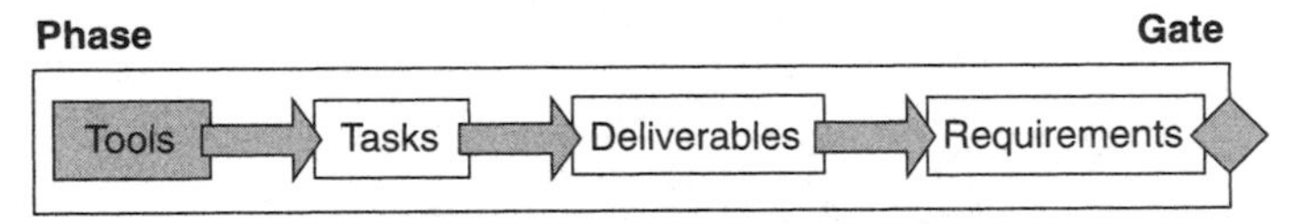

The tool set that supports the Analyze tasks to provide the Analyze step deliverables includes the following methods and best practices:

- Cause-and-effect diagrams
- Process mapping of critical parameters

- Pareto charts
- Affinity diagram (KJ)
- Scatter Plot or Correlation Diagram
- Hypothesis testing
- FMEA

The Improve Step

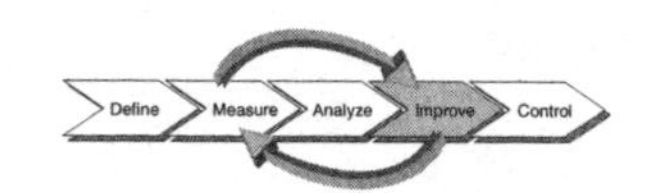

This fourth step in DMAIC involves developing solutions targeted at confirmed causes. The Improve objectives are to verify that the confirmed causes (or critical inputs) are statistically and practically significant and to optimize the process or product/service with the improvements. Sometimes the project goals and objectives are achieved without implementing all the proposed options. Prioritization and validation are important components of selecting a recommended fix to the problem. The project team must quantify the effects of the key variables in the process and develop an improvement plan that modifies the key factors to achieve the desired process improvement.

The four major requirements for the Improve step are as follows:

1. Generate a solution.
2. Select a solution based on agreed-to evaluation criteria.
3. Evaluate the solution for risk.
4. Develop and pilot the implementation path forward.

The final output of this step is to set up for ongoing operations. The path forward plan defines new or "leaned" process steps (if any) and new roles and responsibilities (if any) to execute and maintain the fix on an ongoing basis. Hence, based on the test or pilot information, the path forward plan must begin to determine what controls to put in place to ensure the solution's success in achieving the targets. Then, in the final step, the Control step, the path forward plan is optimized and validated in an operational environment.

Improve Step Tools, Tasks, and Deliverables

The set of tools, tasks, and deliverables included in Figure 7.5 enables the Improve step requirements to be met and ensure measurable results.

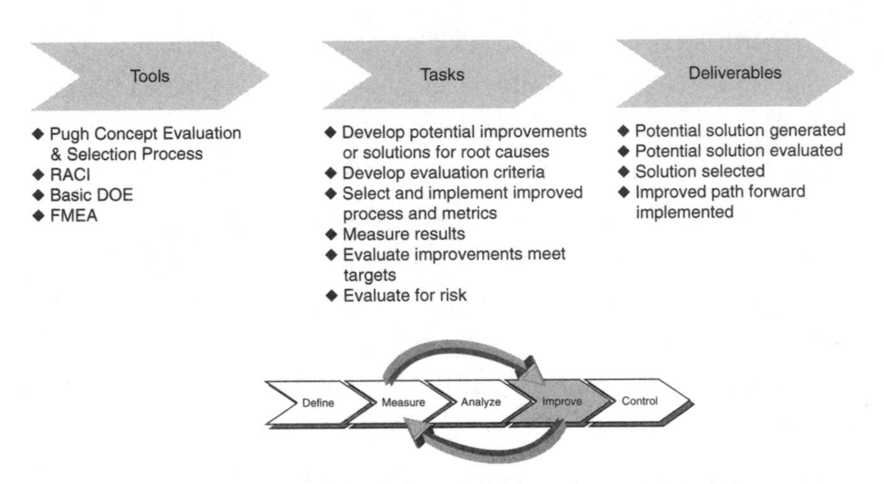

FIGURE 7.5 The tool-task-deliverables combination for the Improve step.

Improve Step Deliverables

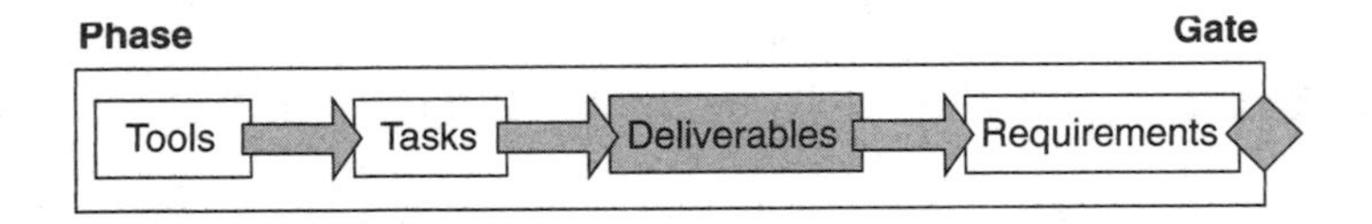

The Deliverables for the Improve step are equally important and should not be shortchanged. The four Improve deliverables are as follows:

1. A potential solution is generated.
2. A potential solution is evaluated.

3. A solution is selected.

4. An improved path forward is implemented.

Improve Step Tasks

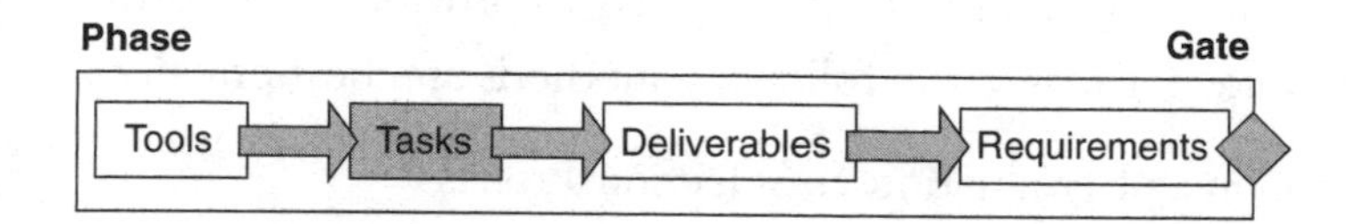

Dictated by the Improve deliverables required to select the solution that best addresses the problem (or opportunity) statement, the short list of tasks within the Improve step includes the following:

1. Develop potential improvements or solutions for root causes:
   a. Create a "validated root cause and proposed solution" chart.
2. Develop evaluation criteria:
   a. Prioritize solution options for each root cause.
   b. Examine solutions with a short-term and long-term approach. Weigh the costs and benefits (or pros and cons) of "quick hit" versus more difficult solution options.
3. Select and implement the improved process and metrics.
4. Measure results.
5. Evaluate whether improvements meet targets.
6. Evaluate for risk.

Tools, Methods, and Best Practices That Enable the Improve Tasks

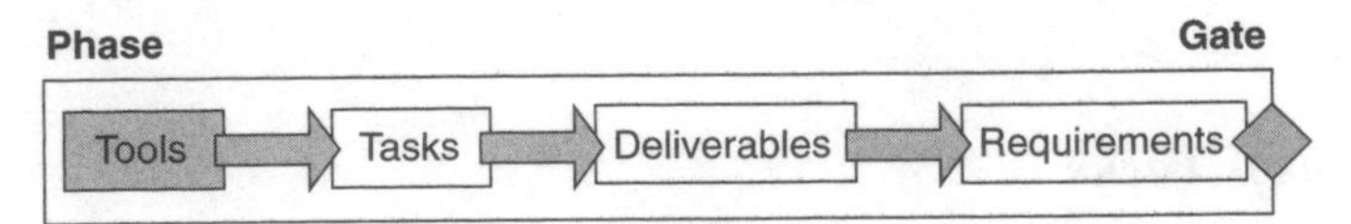

The tool set that supports the Improve tasks to provide the Improve step deliverables consists of the following methods and best practices:

- Pugh Concept Evaluation & Selection Process
- RACI
- Basic DOE
- FMEA

The Control Step

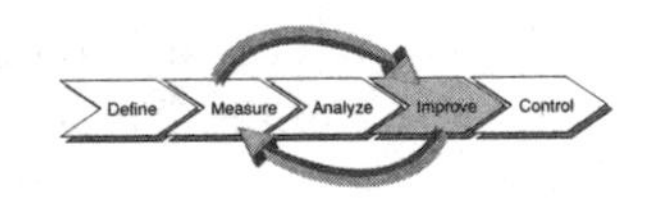

The objective of the fifth and final step in DMAIC is to complete the project work and transition the improved process (or product/ service offering) to the process owner with procedures for maintaining the improvements for ongoing operations. In preparation for the transition, the project team and operations work together to verify the ability to sustain the improvement's long-term capability and plan for continuous process improvement.

The five major requirements for the Control step are as follows:

1. **Standard operating procedures** are documented.
2. **Process control plans** are created or revised.
3. An **established control process** with appropriate metrics and control charts is instituted, preferably in a dashboard format, and a **response plan** is documented and approved.
4. **Improvements are transitioned** to the process owner with supporting documentation.
5. The **project is closed out**, and lessons learned are documented.

Underlying these requirements is the execution of some crucial *supporting plans*, such as change management, the communication plan, the implementation plan, risk management, and the cost/benefit plan. These supporting plans are a part of the transition from the project team to the ongoing management of the solution. Developing these supporting plans should be informed by your best practices and benchmarking.

In concert, these five supporting plans prevent backsliding. Organizations tend to return to their current state—the familiar. Similar to a stretched rubber band, organizations work hard to improve (or stretch) to new, improved shapes, but once you let go, the rubber band snaps back to its original shape. Together, these five plans help sell the project's improvements (or enhancements) and ensure buy-in from management and the process players for the project's long-term goals. Finally, the handoff of responsibilities gains ownership for those who do the day-to-day work (the operational "process players"). They benefit from this final Control step deliverable, because they are the ultimate customers for this step.

Control Step Tools, Tasks, and Deliverables

The set of tools, tasks, and deliverables in Figure 7.6 enables the Control step requirements to be met and ensures measurable results.

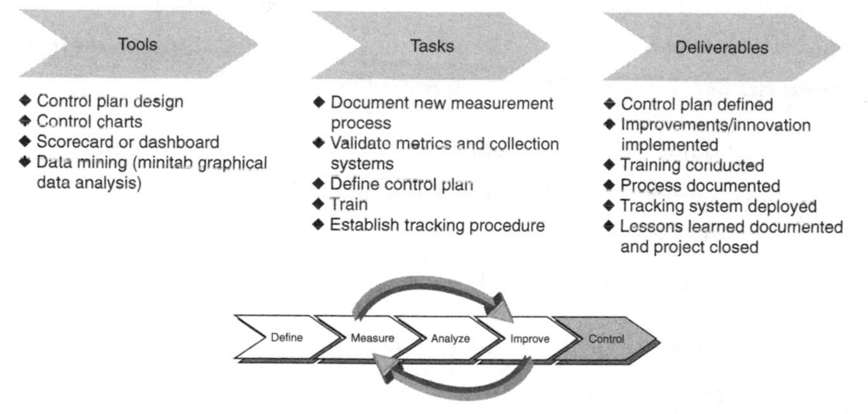

FIGURE 7.6 The tool-task-deliverables combination for the Control step.

Control Step Deliverables

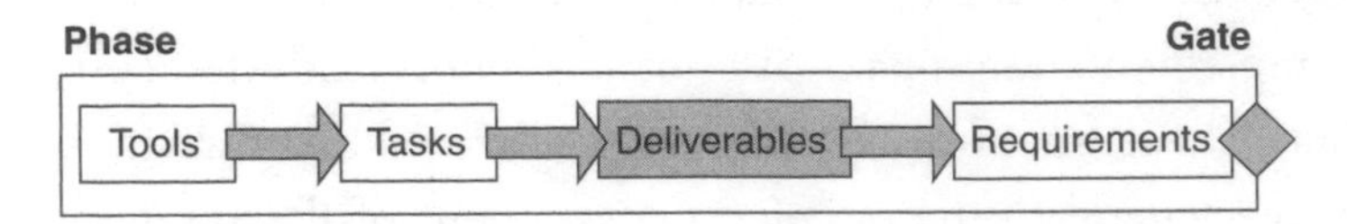

The deliverables for the Control step are equally important and should not be shortchanged. The Control deliverables are as follows:

1. The control plan is defined.
2. Improvements and innovations are implemented.
3. Training is conducted.
4. The process is documented.
5. A tracking system is deployed.
6. Lessons learned are documented and the project is closed.

Control Step Tasks

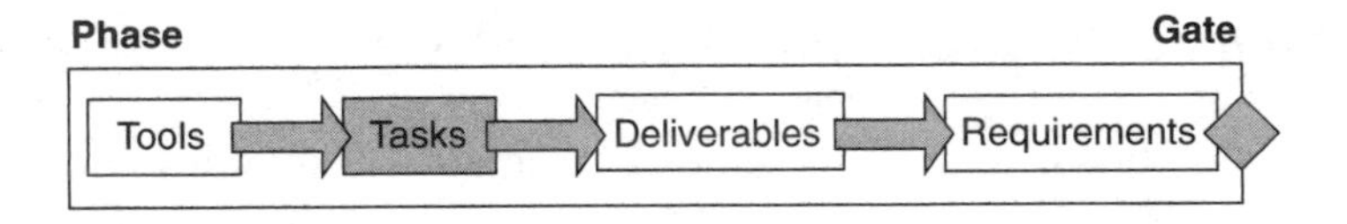

Dictated by the Control deliverables required to select the solution that best addresses the problem (or opportunity) statement, here is the short list of tasks within the Control step:

1. Document a new or improved process and measurements:
   a. Update the detailed process map with any improvements.
   b. Update the RACI Matrix.
2. Validate collection systems and the repeatability and reproducibility of metrics in an operational environment:
   a. Update the Statistical Process Control (SPC) charts.

3. Define the control plan and its supporting plans:
   a. Communications plan of the improvements and operational changes to the customers and stakeholders
   b. Implementation plan
   c. Risk management (and response) plan
   d. Cost/benefit plan
   e. Change management plan
4. Train the operational stakeholders (process owner and players).
5. Establish the tracking procedure in an operational environment:
   a. Monitor implementation.
   b. Validate and stabilize performance gains.
   c. Jointly audit the results and confirm final financials.

Tools, Methods, and Best Practices That Enable the Control Tasks

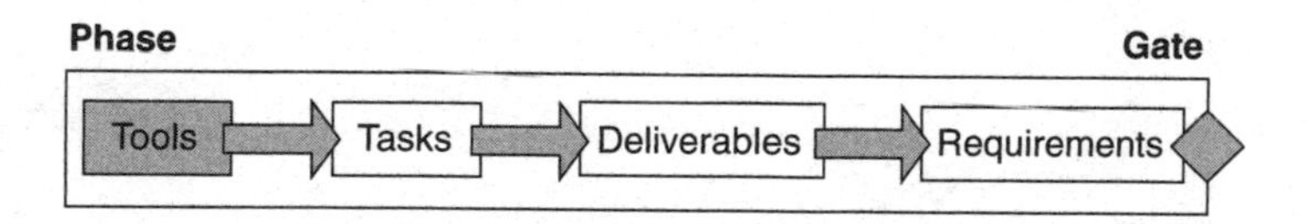

The tool set that supports the Control tasks to provide the Control step deliverables consists of the following methods and best practices:

- Control plan design
- Control charts
- Scorecard or dashboard
- Data mining (for example, Minitab graphical data analysis)

The project team has completed the project and has completed and filed the project close-out documents. The final project review should occur after the improvements demonstrate a steady-state operational achievement of the target. Hence, the review meeting should involve not only the project sponsor and team, but also the key operational stakeholders, to discuss and review lessons learned and to celebrate the project completion.

You now have enhanced the performance of your current process or existing product and services. The Define-Measure-Analyse-Improve-Control approach provides a strong foundation for Six Sigma principles that work in concert with the three Growth approaches described earlier in this book. You can see that the tool set in each approach is similar and is uniquely grouped to specifically answer a particular phase or step requirement. Remember that using *the right tool at the right time* to answer *the right question* is the beauty of following a proven methodology. To unleash the power of Six Sigma—to manage by fact—you need to ensure that you adhere to an established process, wherein its requirements and the tools, tasks, and deliverables are well-defined. You may choose to call your process phases by different names—that's fine. *What you do and what you measure* are what really matter.

Now let's have some fun and look into our crystal ball to see what the future holds.

8

Future Trends in Six Sigma and Marketing Processes

The contribution that "science" provides to marketing is gaining momentum with the advent of improved technology (Customer Relationship Management [CRM] systems, sales automation and work-flow tools, Internet/web/wireless/digital), increased pace of change, and a "flattening" of the world (globalization). The combination of "art" and "science" will continue to be important to overall success, but marketing requires faster access to relevant data to help you make better fact-based decisions. Marketing would benefit from a universal language, structure, and common tool set to better collaborate and innovate not only internally with its colleagues, but also with customers, value chain partners, and "coopetition." ("Coopetition" is the concept wherein competitors join forces in a limited capacity and act as partners to deliver a product or service, often to respond to another third party. This is common in fast-changing industries such as high technology.) The global marketplace will demand better leadership, improved relationships, and more creativity from top firms.

As Lean and Six Sigma mature, we forecast that their impact on business will expand and deepen. Both Lean and Six Sigma can help marketing create a competitive advantage with better information and more proactive management of go-to-market resources and processes and to drive and sustain growth. As a result, we predict that the following business trends will unfold:

- The formation of marketing *centers of excellence* to promote continuous improvement and the standardization and "right-sizing" of marketing tools, tasks, and deliverables.
- Enriched and expanded understanding, on the part of marketing, as to the best practices and methods of executing tools, tasks, and deliverable work flows. It takes time to do things right. As executives who want sustained growth become aligned with the proper structuring (rightsizing) of marketing work, the efficiency and performance of marketing teams will become far more predictable. Critical marketing functions

executed with rigor and discipline produce deliverables with greater certainty.

- Differentiated marketing work flows will be categorized as strategic, tactical, and operational processes similar to those in R&D, product design, and production/service support engineering organizations. The jack-of-all-trades approach will migrate toward more marketing specialization to improve execution excellence that assimilates analytical marketing tools into its best practices.
- Six Sigma concepts will serve as the foundation for a universal marketing language. A greater investment will be made in applying its concepts and implementing specific tool-enabled marketing activities to produce more predictable, successful growth.
- Increased focus on the fundamental marketing variables critical to customer behavioral dynamics ($Y = f(X)$). Better definition of what underwrites a real cause-and-effect relationship, and how to measure and control critical marketing parameters that prevent problems. Measuring variables that signal impending failure rather than measuring failures and reacting to them. Marketing teams will stop measuring what is easy and convenient if it is not fundamental to true cause-and-effect relationships within and across marketing variables.
- A shift from DMAIC Six Sigma for problem-solving and cost control to a phase-gate approach aimed at problem prevention and investment in properly designed marketing work flows to enable growth projects.
- Growing accountability of marketing professionals to drive growth and, with demonstrated success, a rebalancing of marketing and technical/engineering personnel in an enterprise so that marketing can better and more completely perform the

expected tasks. This will reflect the designed balance between an enterprise's marketing and technical innovation strategy.

- Improved collaboration between technical and marketing professionals across strategic, tactical, and operational environments. "That's not my job" attitudes will dissipate and be replaced with better cross-functional teamwork. This will be particularly true in all areas where customer needs, complaints, and sensitivities require translation into technical requirements to avoid future problems by proactive knowledge sharing.
- Transformed thinking to a platform and modular design will help marketing design and monitor the flow of product and service offerings in a balanced portfolio deployment context. An elegant, designed flow of product (and service) families and preplanned line extensions will make it easier to align limited corporate resources to evolving market and competitive dynamics and meet on-time launch demands. This modular approach also reduces the intensity of risk for a single launch and spreads risk across multiple launches.
- Elevated use of an integrated set of scorecards to measure marketing risk to improve decision-making. Checklists must give way to more discriminating scorecards at the marketing tool, task, and deliverable level.

The benefits of applying Lean and Six Sigma to marketing make it worth the investment. Using the Lean and Six Sigma approaches gives decision-makers better information and helps drive uncertainty out of marketing. Companies can better align their product ideas with solid market opportunities and better balance and manage their offerings portfolio by using Lean and Six Sigma to significantly increase the probability of marketplace success. These approaches offer a common structure and language that will facilitate communication throughout the tactical product development and commercialization process between marketing and engineering. Operationally, marketers using Six Sigma become more proactive in managing value-chain resources and go-to-market processes. Scott Fuson, corporate executive director of marketing, sales, and customer service at Dow Corning, gave a compelling testimonial for marketing to embrace Six Sigma. In a recent *Journal of Product Innovation* article, Fuson stated that "The big difference is that before, someone might do something really terrific, but then change jobs or position and the great work would fade away. Six Sigma makes those improvements sustainable; they don't go away...they're built into the process."

We hope you have a better understanding of how to apply Six Sigma and Lean to marketing. It will be a while before Six Sigma for Growth has the same impact on marketing as it has on the design community (using Design for Six Sigma [DFSS]). Your leadership and willingness to take the hard fork in the road will make the difference as you seek sustained profitable growth for your firm.

GLOSSARY

A

affinity diagram

A tool used to gather and group ideas. Usually depicted as a "tree" diagram.

ANOM

Analysis of the Mean. An analytical process that quantifies the mean response for each individual control factor level. ANOM can be performed on data that is in regular engineering units or data that has been transformed into some form of signal-to-noise ratio or another data transform. Main effects and interaction plots are created from ANOM data.

ANOVA

Analysis of the variance. An analytical process that decomposes the contribution each control factor has to the overall experimental response. The ANOVA process also can account for the contribution of interactive effects between control factors and experimental error in the response, provided that enough degrees of freedom are established in the experimental array. The value of epsilon squared (the % contribution to the overall CFR variation) is calculated using data from ANOVA.

array

An arithmetically derived matrix or table of rows and columns that is used to impose an order for efficient experimentation. The rows contain the individual experiments. The columns contain the experimental factors and their individual levels or set points.

ASQ

American Society for Quality.

B

benchmarking

The process of comparative analysis between two or more concepts, components, subassemblies, subsystems, products, or processes. The goal of benchmarking is to qualitatively and quantitatively identify a superior subject within the competing choices. Often the benchmark is used as a standard to meet or surpass. Benchmarks are used to build houses of quality, concept generation, and the Pugh concept selection process.

best practice

A preferred and repeatable action or set of actions completed to fulfill a specific requirement or set of requirements during the phases of a product development process.

beta (ß)

The Greek letter ß is used to represent the slope of a best-fit line. It indicates the linear relationship between the signal factor(s) (Critical Adjustment Parameters) and the measured Critical Functional Response in a dynamic robustness optimization experiment.

Black Belt

A job title or role indicating that the person has been certified as having mastered the Six Sigma DMAIC (Define-Measure-Analyze-Improve-Control) content and has demonstrated expertise in leading one or more projects. This person usually is the team leader of a Six Sigma project, and he or she is often a coach of Green Belts.

blocking

A technique used in classical DOE to remove the effects of unwanted, assignable-cause noise or variability from the experimental response so that only the effects of the control factors are present in the response data. Blocking is a data purification process that helps ensure the integrity of the experimental data used to construct a statistically significant math model.

C

Capability Growth Index (CGI)

The calculated percentage between 0% and 100% that a group of system, subsystem, or subassembly CFRs have attained in getting their Cp indices equal to a value of 2. (This indicates how well their CFRs have attained Six Sigma performance during product development.) The CGI for critical functions is a metric often found on an executive gate review scorecard.

Capability Index
Cp and Cpk indices that calculate the ratio of the Voice of the Customer to the Voice of the Product or process. Cp is a measure of capability based on short-term or small samples of data—usually what is available during product development. Cpk is a measure of long-term or large samples of data that include not only variation of the mean but also the shifting of the mean itself—usually available during steady-state production.

checklist
A simple list of action items, steps, or elements needed to complete a task. Each item is checked off as it is completed.

Classical Design of Experiments (DOE)
Experimental methods employed to construct math models relating a dependent variable (the measured CFR) to the set points of any number of independent variables (the experimental control factors). DOE is used sequentially to build knowledge of fundamental functional relationships (ideal/transfer functions) between various factors and a response variable.

commercialization
A business process that harnesses a company's resources in the endeavor of conceiving, developing, designing, optimizing, certifying design and process capability, producing, selling, distributing, and servicing a product.

compensation
The use of feedforward or feedback control mechanisms to intervene when certain noise effects are present in a product or process. The use of compensation is done only when insensitivity to noise cannot be attained through robustness optimization.

component
A single part in a subassembly, subsystem, or system. An example is a stamped metal part before it has anything assembled to it.

component requirements document
The document that contains all the requirements for a given component. It is often converted into a quality plan that is given to the production supply chain to set the targets and constrain the variation allowed in the incoming components.

control chart
A tool set used to monitor and control a process for variation over time. It varies with the type of data it monitors.

control factor
The factors or parameters (CFP or CTF spec) in a design or process that the engineer can control and specify to define the optimum combination of set points to satisfy the Voice of the Customer.

Critical Adjustment Parameter (CAP)
A specific type of CFP that controls the mean of a CFR. It is identified using sequential DOE and engineering analysis. It is an input parameter for response surface methods for optimizing mean performance after robust design is completed. It allows Cpk to be set equal to Cp, thus allowing entitlement to be approached, if not attained.

Critical Functional Parameter (CFP)
An input variable (usually an engineered additivity grouping) at the subassembly or subsystem level that controls a CFR's mean or variation.

Critical Functional Response (CFR)
A measured scalar or vector (complete, fundamental, continuous engineering variable) output variable that is critical to the fulfillment of a critical (highly important) customer requirement. Some refer to these critical customer requirements as CTQs. A metric often found on an executive gate review scorecard.

criticality
A measurable requirement or functional response that is highly important to a customer. All requirements are important, but only a few are truly critical.

Critical Parameter Management (CPM)
The process that develops critical requirements and measures critical functional responses to design, optimize, and certify a product's capability and its supporting network of manufacturing and service processes.

critical path
The sequence of tasks in a project that takes the greatest amount of time to complete.

Critical-to-Function Specification (CTF)
A dimension, surface, or bulk characteristic (typically a scalar) that is critical to a component's contribution to a subassembly, subsystem, or system-level CFR.

cross-functional team
A group of people representing multiple functional disciplines and possessing a wide variety of technical and experiential backgrounds and skills working together. Particularly applicable in the product commercialization process. (See *multidisciplined team*.)

CTQ
Critical to Quality, as defined by the customer.

D

dashboard
A summary and reporting tool of data and information about a process and/or product performance. Usually viewed as more complex

than a scorecard. Depicts the critical parameters necessary to run the business.

deliverable

A tangible, measurable output completed as an outcome of a task or series of tasks.

Design Capability (Cp_d)

The Cp index for a design's CFR in ratio to its upper and Lower Specification Limits (VOC-based tolerance limits).

Design of Experiments (DOE)

A process for generating data that uses a mathematically derived matrix to methodically gather and evaluate the effect of numerous parameters on a response variable. Designed experiments, when properly used, efficiently produce useful data for model building or engineering optimization activities.

deterioration noise factor

A source of variability that results in some form of physical deterioration or degradation of a product or process. This is also called an "inner noise" because it refers to variation inside the controllable factor levels.

DFSS

Design for Six Sigma. A Six Sigma concept used by the engineering technical community to design and develop a product.

DMADV

Define-Measure-Analyze-Design-Validate. A five-step Six Sigma method used primarily to redesign a broken process, as well as to solve problems and/or improve processes or products.

DMAIC

Define-Measure-Analyze-Improve-Control. A five-step Six Sigma method used to solve problems and/or improve processes or products.

DMEDI

Define-Measure-Explore-Develop-Implement. A five-step method combining classic Six Sigma and Lean concepts to redesign a broken process, as well as to solve problems and/or improve processes or products.

DPMO

Defects Per Million Opportunities measurement.

E

economic coefficient

Used in the quality loss function. It represents the proportionality constant in the loss function of the average dollars lost (A_0) due to a customer reaction to off-target performance and the square of the deviation from the target response ($\varnothing_0^2$). This is typically, but not exclusively, calculated when approximately 50% of the customers are motivated to take some course of economic action due to poor performance (but not necessarily outright functional failure). This is often referred to as the LD 50 point in the literature.

ECV
Expected Commercial Value. A financial metric often found on an executive gate review scorecard.

energy flow map
A representation of an engineering system that shows the paths of energy divided into productive and nonproductive work. Analogous to a free-body diagram from an energy perspective. This accounts for the law of conservation of energy and is used in preparation for math modeling and Design of Experiments.

energy transformation
The physical process that a design or product system uses to convert some form of input energy into various other forms of energy that ultimately produce a measurable response. The measurable response may itself be a form of energy or the consequence of energy transformations that have taken place within the design.

engineering metric
A scalar or vector that is usually called a CTF spec, CFP, CAP, or CFR. It is greatly preferred over quality metrics (such as yield and defects) in DFSS.

engineering process
A set of disciplined, planned, and interrelated activities that engineers employ to conceive, develop, design, optimize, and certify the capability of a new product or process design.

environmental noise factor
A source of variability that is due to effects that are external to the design or product. Also called "outer noise." This can also be a source of variability that one neighboring subsystem imposes on another neighboring subsystem or component. Examples include vibration, heat, contamination, misuse, and overloading.

experiment
An evaluation or series of evaluations that explore, define, quantify, and build data that can be used to model or predict functional performance in a component, subassembly, subsystem, or product. Experiments can be used to build fundamental knowledge for scientific research. They also can be used to design and optimize product or process performance in the engineering context of a specific commercialization process.

experimental efficiency

A process-related activity that is facilitated by intelligent application of engineering knowledge and the proper use of designed experimental techniques. Examples include the use of fractional factorial arrays, control factors that are engineered for additivity, and compounded noise factors.

experimental error

The variability present in experimental data that is caused by meter error and drift, human inconsistency in taking data, random variability taking place in numerous noise factors not included in the noise array, and control factors that have not been included in the inner array. In the Taguchi approach, variability in the data due to interactive effects is often but not always included as experimental error.

experimental factor

An independent parameter that is studied in an orthogonal array experiment. Robust design classifies experimental factors as either control factors or noise factors.

experimental space

The combination of all the control factor, noise factor, and signal factor (CAP) levels that produce the range of measured response values in an experiment.

F

feedback control system

A method of compensating for the variability in a process or product by sampling output response and sending a feedback signal that changes a CAP to put the response's mean back on its intended target.

FMEA

Failure Modes & Effects Analysis. A risk analysis technique that identifies and ranks the potential failure modes of a design or process and then prioritizes improvement actions.

F-Ratio

The ratio formed in the ANOVA process by dividing the mean square of each experimental factor effect by the mean square of the error variance. This is the ratio of variation that occurs *between* each of the experimental factors in comparison to the variation occurring *within all the experimental factors* being evaluated in the entire experiment. It is a form of signal-to-noise ratio in a statistical sense. The noise in this case is random experimental error—not variability due to the assignable-cause noise factors in the Taguchi noise array.

Full Factorial Design

Two- and three-level orthogonal arrays that include every possible combination between the experimental factors. Full factorial experimental designs use degrees of freedom to account for all the main effects and all interactions between factors included in the experimental array. Basically, all interactions beyond two-way interactions are likely to be of negligible consequence. Thus, there is little need to use large arrays to rigorously evaluate such rare and unlikely three-way interactions and above.

fundamental
The property of a CFR that expresses the basic or elemental physical activity that is ultimately responsible for delivering customer satisfaction. A response is fundamental if it does not mix mechanisms and is uninfluenced by factors outside the component, subassembly, subsystem, and system design or production process being optimized.

G

Gantt Chart
A horizontal bar chart used for project planning and control. Lists the necessary project activities as row headings against horizontal lines showing the dates and duration of each activity.

gate
A short period of time during a process when the team reviews and reacts to the results against requirements from the previous phase and proactively plans for the smooth execution of the next phase.

gate review
Meeting with the project team and sponsors to inspect completed deliverables. You focus on the results from specific tools and best practices, and manage the associated risks and problems. You also make sure that the team has everything it needs to apply the tools and best practices for the next phase with discipline and rigor. A gate review's time should be 20% reactive and 80% proactive.

goalpost mentality
A philosophy that uses a sports analogy, about quality that accepts anything within the tolerance band (USL-LSL) as equally good and anything that falls outside the tolerance band as equally bad.

goal statement
Identifies the critical parameters (including time frame) for a targeted improvement. (Use the SMART technique to ensure completeness.)

GOSPA
Goals, Objectives, Strategies, Plans, and Actions planning methodology.

grand Total Sum of Squares
The value obtained by squaring the response of each experimental run from a matrix experiment and then adding the squared terms together.

Green Belt
A job title or role indicating that the person has been certified as having demonstrated an understanding of the basic Six Sigma DMAIC concepts. This person may support a Black Belt on a Six Sigma project. In some companies, this person works on small-scale projects directly related to his or her job.

H

histogram
A graphical display of the frequency distribution of a set of data. Histograms display the shape, dispersion, and central tendency of a data set's distribution.

House of Quality
An input/output relationship matrix used in the process of QFD.

hypothesis testing
A statistical evaluation that checks a statement's validity to a specified degree of certainty. These tests are done using well-known and quantified statistical distributions.

I

IDEA
Identify-Define-Evaluate-Activate. A four-step Six Sigma method used by strategic marketing to define, develop, manage, and refresh a portfolio of offerings (products and services).

ideal/transfer function
A fundamental functional relationship between various engineering control factors and a measured Critical Functional Response variable. The math model of $Y = f(x)$ that represents the customer-focused response that would be measured if no noise or only random noise were acting on the design or process.

inbound marketing
Marketing activities that are focused on providing deliverables for internal consumption, as opposed to deliverables intended for the marketplace.

independent effect
The nature of an experimental factor's effect on the measured response when it is acting independently with respect to any other experimental factor. When all control factors produce independent effects, the design is said to be exhibiting an additive response.

inference
Drawing some form of conclusion about a measurable functional response based on representative or sample experimental data. Sample size, uncertainty, and the laws of probability play a major role in making inferences.

inner array

An orthogonal matrix that is used for the control factors in a designed experiment and that is crossed with some form of outer noise array during robust design.

inspection

The process of examining a component, subassembly, subsystem, or product for off-target performance, variability, and defects during product development or manufacturing. The focus typically is on whether the item under inspection is within the allowable tolerances. Like all processes, inspection itself is subject to variability. Out-of-spec parts or functions may pass inspection inadvertently.

interaction

The dependence of one experimental factor on the level set point of another experimental factor for its contribution to the measured response. The two types of interaction are synergistic (mild to moderate and useful in its effect) and antisynergistic (strong and disruptive in its effect).

interaction graph

A plot of the interactive relationship between two experimental factors as they affect a measured response. The ordinate (vertical axis) represents the response being measured, and the abscissa (horizontal axis) represents one of the two factors being evaluated. The average response value for the various combinations of the two experimental factors is plotted. The points representing the second factor's low level are connected by a line. Similarly, the points representing the second factor's next-higher level are connected by a line.

IRR

Internal Rate of Return (IRR). A financial metric often found on an executive gate review scorecard.

K

KJ Analysis

Jiro Kawakita was a Japanese anthropologist who treated attribute (or language) data similar to variable data by grouping and prioritizing it. A KJ diagram (similar to an affinity diagram) focuses on the unique and different output, linking the critical customer priorities to the project team's understanding and consensus.

L

lagging indicator

An indicator that follows the occurrence of something. Hence, it is used to determine the performance of an occurrence or an event. By tracking lagging indicators, you react to the results. Examples are the high and low temperatures, precipitation, and humidity for a given day.

leading indicator

An indicator that precedes the occurrence of something. Hence, it is used to signal the upcoming occurrence of an event. By tracking leading indicators, you can prepare for or anticipate the subsequent event and be proactive. For example, barometric pressure and Doppler radar for the surrounding region indicate the upcoming weather.

Lean Six Sigma

A modified Six Sigma approach that emphasizes improving a process's speed by "leaning" it of its non-value-adding steps. Typically used in a manufacturing environment. Its common metrics include zero wait time, zero inventory, line balancing, cutting batch sizes to improve flow-through, and reducing overall process time.

level

The set point where a control factor, signal factor (CAP), or noise factor is placed during a designed experiment.

level average analysis

See ANOM.

life-cycle cost

The costs associated with making, supporting, and servicing a product or process throughout its intended life.

linear combination

This term has a general mathematical definition and a specific mathematical definition associated with the dynamic robustness case. In general, a linear combination is the simple summation of terms. In the dynamic case, it is the specific summation of the product of the signal level and its corresponding response ($M_i yi_{i,j}$).

linear graph

A graphical aid used to assign experimental factors to specific columns when evaluating or avoiding specific interactions.

linearity

The relationship between a dependent variable (the response) and an independent variable (the signal or control factor) that is graphically expressed as a straight line. Linearity typically is a topic within the dynamic cases of the robustness process and in linear regression analysis.

LMAD

Launch-Manage-Adapt-Discontinue. A four-step Six Sigma method used by marketing to manage the ongoing operations of a portfolio of launched offerings (products and services) across the value chain.

loss to society

The economic loss that society incurs when a product's functional performance deviates from its targeted value. This loss is often due to economic action taken by the consumer reacting to poor product performance. It can also be due to the effects that spread through society when products fail to perform as expected. For example, a new car breaks down in a busy intersection because of a transmission defect, causing 14 people to be 15 minutes late for work (cascading loss to many points in society).

Lower Specification Limit (LSL)
The lowest functional performance set point that a design or component can attain before functional performance is considered unacceptable.

M

main effect
The contribution that an experimental factor makes to the measured response independent of experimental error and interactive effects. The sum of the half effects for a factor is equal to the main effect.

manufacturing process capability (Cp_m)
The ratio of the manufacturing tolerances to the measured performance of the manufacturing process.

matrix
An array of experimental set points that is derived mathematically. Composed of rows (containing experimental runs) and columns (containing experimental factors).

matrix experiment
A series of evaluations that are conducted under the constraint of a matrix.

mean
The average value of a sample of data that is typically gathered in a matrix experiment.

Mean Square Deviation (MSD)
A mathematical calculation that quantifies the average variation a response has with respect to a target value.

mean square error
A mathematical calculation that quantifies the variance within a set of data.

measured response
The quality characteristic that is a direct measure of functional performance.

measurement error
The variability in a data set that is due to poorly calibrated meters and transducers; human error in reading and recording data; and normal, random effects that exist in any measurement system used to quantify data.

meter
A measurement device usually connected to some sort of transducer. The meter supplies a numerical value to quantify functional performance.

Monte Carlo Simulation
A computer simulation technique that uses sampling from a random-number sequence to simulate characteristics, events, or outcomes with multiple possible values.

MSA
Measurement System Analysis. A tool that helps you understand the level of reproducibility and repeatability.

MTBF
Mean Time Between Failure. A measurement of the lapsed time from one failure to the next.

multidisciplined team
A group of people possessing a wide variety of technical and experiential backgrounds and skills, working together. Particularly applicable in the product commercialization process. (See *cross-functional team.*)

N

noise
Any source of variability. Typically noise is external to the product (such as environmental effects), is a function of unit-to-unit variability due to manufacturing, or is associated with the effects of deterioration. In this context, noise is an assignable, nonrandom cause of variation.

noise directionality
A distinct upward or downward trend in the measured response, depending on the level at which the noises are set. Noise factor set points can be compounded, depending on the response's directional effect.

noise experiment
An experiment designed to evaluate the strength and directionality of noise factors on a product or process response.

noise factor
Any factor that promotes variability in a product or process.

Nominal-the-Best (NTB)
A case in which a product or process has a specific nominal or targeted value.

normal distribution
The symmetric distribution of data about an average point. A normal distribution takes the form of a bell-shaped curve. It is a graphic illustration of how randomly selected data points from a product or process response

mostly fall close to the average response, with fewer and fewer data points falling farther and farther away from the mean. The normal distribution can also be expressed as a mathematical function and is often called a Gaussian distribution.

NPV
Net Present Value. A financial metric often found on an executive gate review scorecard.

NUD
New, Unique, and Difficult.

O

offline quality control
The processes included in preproduction commercialization activities. The processes of concept design, parameter design, and tolerance design make up the elements of offline quality control. It is often viewed as the area where quality is designed into the product or process.

one-factor-at-a-time experiment
An experimental technique that examines one factor at a time, determines the best operational set point, locks in on that factor level, and then moves on to repeat the process for the remaining factors. This technique is widely practiced in scientific circles but lacks the circumspection and discipline provided by full and fractional factorial designed experimentation. Sometimes one-factor-at-a-time experiments are used to build knowledge before a formal factorial experiment is designed.

online quality control
The processes included in the production phase of commercialization. The processes of Statistical Process Control (loss-function-based and traditional), inspection, and evolutionary operation (EVOP) are examples of online quality control.

operating income
Calculated as gross profit minus operating expenses. A financial metric often found on an executive gate review scorecard.

operational marketing
Pertains to marketing's activities in support of launching and managing an offering (product and/or service) or set of offerings across the value chain.

optimize
Finding and setting control factor levels at the point where their mean, standard deviation, or signal-to-noise ratios are at the desired or maximum value. Optimized performance means that the control factors are set such

that the design is least sensitive to the effects of noise and the mean is adjusted to be right on the desired target.

orthogonal
The property of an array or matrix that gives it balance and lets it produce data that allows for the independent quantification of independent or interactive factor effects.

orthogonal array
A balanced matrix that is used to lay out an experimental plan for the purpose of designing functional performance quality into a product or process early in the commercialization process.

outbound marketing
Marketing activities that are focused on providing deliverables for the customers as opposed to deliverables intended for internal consumption.

outer array
The orthogonal array used in dynamic robust design that contains the noise factors and signal factors. Each treatment combination of the control factors specified in the inner array is repeated using each of the treatment combinations specified by the outer array.

P

parameter
A factor used in the design, optimization, and certification of capability processes. Experimental parameters are CFRs, CFPs, CTF specs, and noise factors.

parameter design
The process employed to optimize the levels of control factors against the effect of noise factors. Signal factors (dynamic cases) or tuning factors (NTB cases) are used in the two-step optimization process to adjust the performance to a specific target during parameter (robust) design.

parameter optimization experiment
The main experiment in parameter design that is used to find the optimum level for the control factors. Usually this experiment is done using some form of dynamic crossed array design.

Pert Chart
A diagram that displays the dependency relationships that exist between tasks.

phase
A period of time in which you conduct work to produce specific results that meet the requirements for a given project, wherein specific tools and best practices are used.

phase/gate product development process
A series of time periods that are rationally divided into phases for the development of new products and processes. Gates are checkpoints at the end of each phase to review progress, assess risks, and plan for efficiency in future phase performance.

population parameter or statistic
A statistic such as the mean or standard deviation that is calculated with all the possible values that make up the entire population of data in an experiment. Samples are just a portion of a population.

Porter's 5 Forces Analysis
A business analysis model, developed by Michael Porter, to determine an industry's competitive nature by evaluating five forces: Rivalry among Current Players, Barriers to Entry, Threat of Substitutes, Buyer (Bargaining) Power, and Supplier Bargaining Power.

probability
The likelihood or chance that an event or response will occur out of some number (n) of possible opportunities.

problem statement or opportunity for improvement
A clear, concise definition of what is wrong with a current process or product/services offering. Should be aligned with a company's strategies and/or annual plan. (Use the SMART technique to ensure completeness.)

process
A set sequence of steps to make something or do something.

Process Capability Analysis
Quantifies a process's ability to produce output that meets customer requirements. Various capability metrics include DPMO (Defects Per Million Opportunities); Cp, Cpk (potential process capability, short-term); Pp, Ppk (process capability, long-term); Rolled Throughput Yield (RTY).

process map
A type of flow chart depicting the steps in a process and its inputs and outputs. It often identifies responsibility for each step and the key measures.

product commercialization
The act of gathering customer needs; defining requirements; conceiving product concepts; selecting the best concept; and designing, optimizing, and certifying the capability of the superior product for production, delivery, sales, and service.

product development
The continuum of tasks, from inbound marketing, to technology development, to certified technology being transferred into product

design, to the final step of the certified product design being transferred into production.

project cycle time
The time that elapses from the beginning to the end of a project.

project management
The methods of planning, designing, managing, and completing projects. Project management designs and controls the micro timing of tasks and actions (underwritten by tools and best practices) within each phase of the product development process.

Pugh Process
A structured concept selection process used by multidisciplinary teams to converge on superior concepts. This process uses a matrix consisting of criteria based on the Voice of the Customer and its relationship to specific candidate design concepts. The evaluations are made by comparing the new concepts to a benchmark called the datum. The process uses the classification metrics of "same as the datum," "better than the datum," or "worse than the datum." Several iterations are employed wherever increasing superiority is developed by combining the best features of highly ranked concepts until a superior concept emerges and becomes the new benchmark.

Q

QFD
Quality Function Deployment. A process for translating the Voice of the Customer into technical requirements at the product level. QFD, as part of the Critical Parameter Management process, uses a series of matrices called "houses of quality" to translate and link system requirements to subsystem requirements. Those, in turn, are translated and linked to subassembly requirements. Those, in turn, are translated and linked to component requirements. Those, in turn, are translated and linked to manufacturing process requirements.

quality
The degree or grade of excellence. In a product development context, it is a product with superior features that performs on-target with low variability throughout its intended life. In an economic context, it is the absence or minimization of costs associated with the purchase and use of a product or process.

quality characteristic
A measured response that relates to a general or specific requirement that can be an attribute or a continuous variable. The quantifiable measure of performance that directly affects the customer's satisfaction.

Often in DFSS these have to be converted to an engineering scalar or vector.

quality engineering
Most often referred to as Taguchi's approach to offline quality control (concept, parameter, and tolerance design) and online quality control.

Quality Function Deployment (QFD)
A disciplined process for obtaining, translating, and deploying the Voice of the Customer into the various phases of technology development and the ensuing commercialization of products or processes during product design.

quality loss cost
The costs associated with the loss to customers and society when a product or process performs off the targeted response.

quality loss function
The relationship between the dollars lost by a customer due to off-target product performance and the product's measured deviation from its intended performance. Usually described by the quadratic loss function.

quality metric
Defects, time-to-failure, yield, go/no go. (See *quality characteristic*.)

quality plan
The document that is used to communicate specifications to the production supply chain. Often the component House of Quality is converted into a quality plan.

R

RACI Matrix
A two-dimensional table that lists tasks or deliverables as the row headings and roles (or people's names) as the column headings. The cell data contains the responsibility, by task, by role (or person): R = responsible, A = accountable, C = consulted, and I = informed.

random error
The nonsystematic variability that is present in experimental data due to random effects occurring outside the control factor main effects. The residual variation in a data set that is induced by unsuppressed noise factors and error due to human or meter error.

randomization
The technique employed to remove or suppress the effect of systematic or biased order (a form of assignable-cause variation) in running designed experiments. Randomizing is especially important when applying classic DOE in the construction of math models. It helps ensure that the data is as random as possible.

Real-Win-Worth (RWW) Analysis
A technique to analyze market potential, competitive position, and financial return. The composite value is often used as a summary metric on an executive gate review scorecard.

relational database
The set of requirements and fulfilling data that is developed and documented during Critical Parameter Management. It links many-to-many relationships throughout the hierarchy of the system being developed.

reliability
The measure of robustness over time. How long a product or process performs as intended.

repeatability
Variation of repeated measurements of the *same item.*

repeat measurement
The taking of data points where the multiple measured responses are taken without changing any of the experimental set points. Repeat measurements provide an estimate of measurement error only.

replicate
The taking of data in which the design or process set points have all been changed since the previous readings were taken. Often a replicate is taken for the first experimental run and then again at the middle and end of an experiment (for a total of three replicates of the first experimental run). Replicate measurements provide an estimate of total experimental error.

reproducibility
1. The variation in the averages from repeated measurements made by *different people* on *the same item.* 2. The ability of a design to perform as targeted throughout the entire development, design, and production phases of commercialization. Verification tests provide the data on reproducibility in light of the noise imposed on the design.

requirement
Criteria that must be fulfilled; something wanted or needed.

response
The measured value taken during an experimental run. Also called the quality characteristic. In DFSS we prefer to focus on Critical Functional Responses (CFRs).

risk mitigation
A planning process to identify, prevent, remove, or reduce risk if it occurs and to define actions to limit the severity and impact of a risk should it occur.

robust design
A process within the domain of quality engineering for making a product or process insensitive to the effects of variability without actually removing the sources of variability. Synonymous with parameter design.

ROI
Return on Investment. Calculated as the annual benefit divided by the investment amount. A financial metric often found on an executive gate review scorecard.

S

sample
A select, usually random set of data points that are taken out of a greater population of data.

sample size
The measure of how many samples have been taken from a larger population. Sample size has a notable effect on the validity of making a statistical inference.

sample statistic
A statistic such as the mean or standard deviation that is calculated using a sample from the values that make up the entire population of data.

saturated experiment
The complete loading of an orthogonal array with experimental factors. There is the possibility of confounding main effects with potential interactions between the experimental factors within a saturated experiment.

scalar
A continuous engineering variable that is measured by its magnitude alone (no directional component exists for a scalar).

scaling factor
A critical adjustment parameter that is known to have a strong effect on the mean response and a weak effect on the standard deviation. The scaling factor often has the additional property of possessing a proportional relationship to both the standard deviation and the mean.

scaling penalty
The inflation of the standard deviation in a response as the mean is adjusted using a CAP.

scorecard
A set of critical summary data used to predict outcomes or evaluate performance of a process or product when making decisions and managing risk. Often called a "dashboard."

screening experiment
Typically a small, limited experiment that is used to determine which factors are important to the response of a product or process. Screening experiments are used to build knowledge before the main modeling experiments in sequential DOE methodology.

sensitivity
The change in a CFR based on unit changes in a CFP, CAP, or CTF spec. Also a measure of the magnitude (steepness) of the slope between the measured response and the signal factor (CAP) levels in a dynamic robustness experiment.

sigma (σ)
The standard deviation (technically a measure of the population standard deviation).

signal factor
A CAP that is known to be capable of adjusting the design's average output response in a linear manner. Signal factors are used in dynamic robustness cases as well as in response surface methods.

Signal-to-Noise Ratio
A ratio or value formed by transforming the response data from a robust design experiment using a logarithm to help make the data more additive. Classically, signal-to-noise is an expression relating the useful part of the response to the nonuseful variation in the response.

SIPOC
A summary tool to communicate the Suppliers, Inputs, Process, Outputs, and Customers of a process.

Six Sigma (6s)
A disciplined approach to enterprise-wide quality improvement and variation reduction. Technically, it is the denominator of the capability (Cp) index.

slope
Quantifies the linear relationship between the measured response and the signal factor (CAP). (See *beta*.)

smaller-the-better (STB)
A static case where the smaller the measured response is, the better the quality of the product or process.

SMART problem and goal statement
An acronym specifying that a project problem and goal statement need to be Specific, Measurable, Achievable (but aggressive), Relevant (to the project team and business), and Time-bounded.

SPC
Statistical Process Control.

specification
A specific quantifiable set point that typically has a nominal or target value and a tolerance of acceptable values associated with it. They are what results when a team tries to use tools and best practices to fulfill a requirement. Ideally, requirements would be completely fulfilled by a final design specification, but many times they are not.

sponsor
A role indicating that the person has ultimate accountability for a Six Sigma project, its direction, and its funding. This person is a senior manager whose job is highly dependent on the project's outcome. Often this person is the functional process owner of the process on which the project is focused. The project team reports to the sponsor for this project. The sponsor conducts Gate Reviews of the project and serves as the project's liaison to the company's goals and missions.

standard deviation
A measure of the variability in a set of data. It is calculated by taking the square root of the variance. Standard deviations are not additive, but the variances are.

static robustness case
One of the two major types of experimental robustness cases to study a product or process response as related to specific design parameters. The static case has no predesigned signal factor associated with the response. Thus, the response target is considered fixed or static. Control factors and noise factors are used to find local optimum set points for static robustness performance.

strategic marketing
Pertains to marketing's activities in support of the definition, development, management, and refresh of a portfolio of offerings (products or services) at an enterprise, business unit, or division level.

subassembly
Any two or more components that can be assembled into a functioning assembly.

subassembly requirements document
A document that contains the requirement, both critical and all others, for a subassembly.

subsystem
A group of individual components and subassemblies that perform a specific function within the total product system. Systems consist of two or more subsystems.

subsystem requirements document
A document that contains the requirement, both critical and all others, for a subsystem.

Sum of Squares
A calculation technique used in the ANOVA process to help quantify the effects of the experimental factors and the mean square error (if replicates have been taken).

Sum of Squares due to the mean
The calculation of the Sum of Squares to quantify the overall mean effect due to the experimental factors being examined.

supply chain
The network of suppliers that provide raw materials, components, subassemblies, subsystems, software, or complete systems to your company.

surrogate noise factor
Time, position, and location are not actual noise factors but stand in nicely as surrogate "sources of noise" in experiments that do not have clearly defined physical noises. These are typically used in process robustness optimization cases.

synergistic interaction
A mild to moderate form of interactivity between control factors. Synergistic interactions display monotonic but nonparallel relationships between two control factors when their levels are changed. They typically are not disruptive to robust design experiments.

system
An integrated group of subsystems, subassemblies, and components that make up a functioning unit that harmonizes the mass, energy, and information flows and transformations of the elements to provide an overall product output that fulfills the customer-based requirements.

system integration
The construction and evaluation of the system from its subsystems, subassemblies, and components.

system requirements document
A document that contains the requirements, both critical and all others, for a system.

T

tactical marketing
Pertains to marketing's activities in support of the design and development of an offering (product or service) to prepare it for commercialization or launch.

Taguchi, Genichi
The originator of the well-known system of quality engineering. Dr. Taguchi is an engineer, former university professor, author, and global quality consultant.

target
The ideal point of performance that is known to provide the ultimate in customer satisfaction. Often called the "nominal set point" or the "ideal performance specification."

task
A specific piece of work (or definable activity) required to be done as a duty.

technology development
The building of new or leveraged technology (research and development) in preparation for transfer of certified technology into product design programs.

technology transfer
The handing over of certified (robust and tunable) technology and data acquisition systems to the design organization.

tool
An instrument or device that helps you complete a task.

Total Sum of Squares
The part of the ANOVA process that calculates the Sum of Squares due to the combined experimental factor effects and the experimental error. It is the Total Sum of Squares that is decomposed into the Sum of Squares due to individual experimental factor effects and the Sum of Squares due to experimental error.

transfer function
The fundamental functional relationship between various engineering control factors and a measured Critical Functional Response variable. The math model of $Y = f(x)$ that represents the customer-focused response that would be measured if there were no noise or only random noise acting on the design or process. Sometimes this is called a "transfer function" because it helps model how energy, mass, and logic and control signals are transferred across system, subsystem, and subassembly boundaries.

treatment combination
A single experimental run from an orthogonal array.

TRIZ
A systematic innovation process originally developed in Russia to generate ingenious and viable ideas.

two-step optimization process
The process of first finding the optimum control factor set points to minimize sensitivity to noise and then adjusting the mean response to the customer-focused target.

U

UAPL
Understand-Analyze-Plan-Launch. A four-step Six Sigma method used by tactical marketing to ready an offering (products and services) for commercialization. This process is designed to be conducted in parallel with the technical community developing a product or service.

UMC
Unit Manufacturing Cost. The cost associated with making a product or process.

unit-to-unit variability
Variation in a product or process due to noises in the production and assembly process.

Upper Specification Limit (USL)
The largest functional performance set point that a design or component is allowed before functional performance is considered unacceptable.

V

value chain
For a launched offer, the go-to-market value chain describes functions that add value (according to the customer or client) to an already launched offering (product or service). Examples include sales, value-added resellers, other distribution channels, consultants (delivering incremental software and services), service (break-fix, maintenance, parts), training, customer support call center, and marketing support materials and advertising.

variance
The mean squared deviation of the measured response values from their average value.

variation
Changes in parameter values due to systematic or random effects. Variation is the root cause of poor quality and the monetary losses associated with it.

vector
An engineering measure that has both magnitude and directionality associated with it. These are highly valued metrics for CPM.

verification
The process of validating the results from a model or designed experiment.

Voice of the Business (VOB)
The business's internal requirements in the words of senior or executive management. The VOB is used throughout the product commercialization process, and it must be balanced with customer requirements.

Voice of the Customer (VOC)
The customer's wants and needs in his or her own words. The VOC is used throughout the product commercialization process to keep the requirements and the designs that fulfill them focused on the customer's needs.

Voice of the Process
The measured functional output of a manufacturing process.

Voice of the Product
The measured functional output of a design element at any level within the engineered system.

W

WCBF
Worldwide Conventions and Business Forums.

Work Breakdown Structure (WBS)
The process of dividing a project into manageable tasks and sequencing them to ensure a logical flow between tasks.

work-flow chart
A diagram that depicts the flow of work in a process.

Y

Y = f(X)
A mathematical equation read as "Y equals f of X." This means that the result (output) measures (represented by Y) are a function of the process (input) measures (represented by X). Used to understand how the inputs affect the outputs.

yield
The percentage of the total number of units produced that are acceptable; percent good.

INDEX

THIS BOOK IS SAFARI ENABLED

INCLUDES FREE 45-DAY ACCESS TO THE ONLINE EDITION

The Safari® Enabled icon on the cover of your favorite technology book means the book is available through Safari Bookshelf. When you buy this book, you get free access to the online edition for 45 days.

Safari Bookshelf is an electronic reference library that lets you easily search thousands of technical books, find code samples, download chapters, and access technical information whenever and wherever you need it.

TO GAIN 45-DAY SAFARI ENABLED ACCESS TO THIS BOOK:

- Go to **http://www.phptr.com/safarienabled**
- Complete the brief registration form
- Enter the coupon code found in the front of this book on the "Copyright" page

If you have difficulty registering on Safari Bookshelf or accessing the online edition, please e-mail customer-service@safaribooksonline.com.

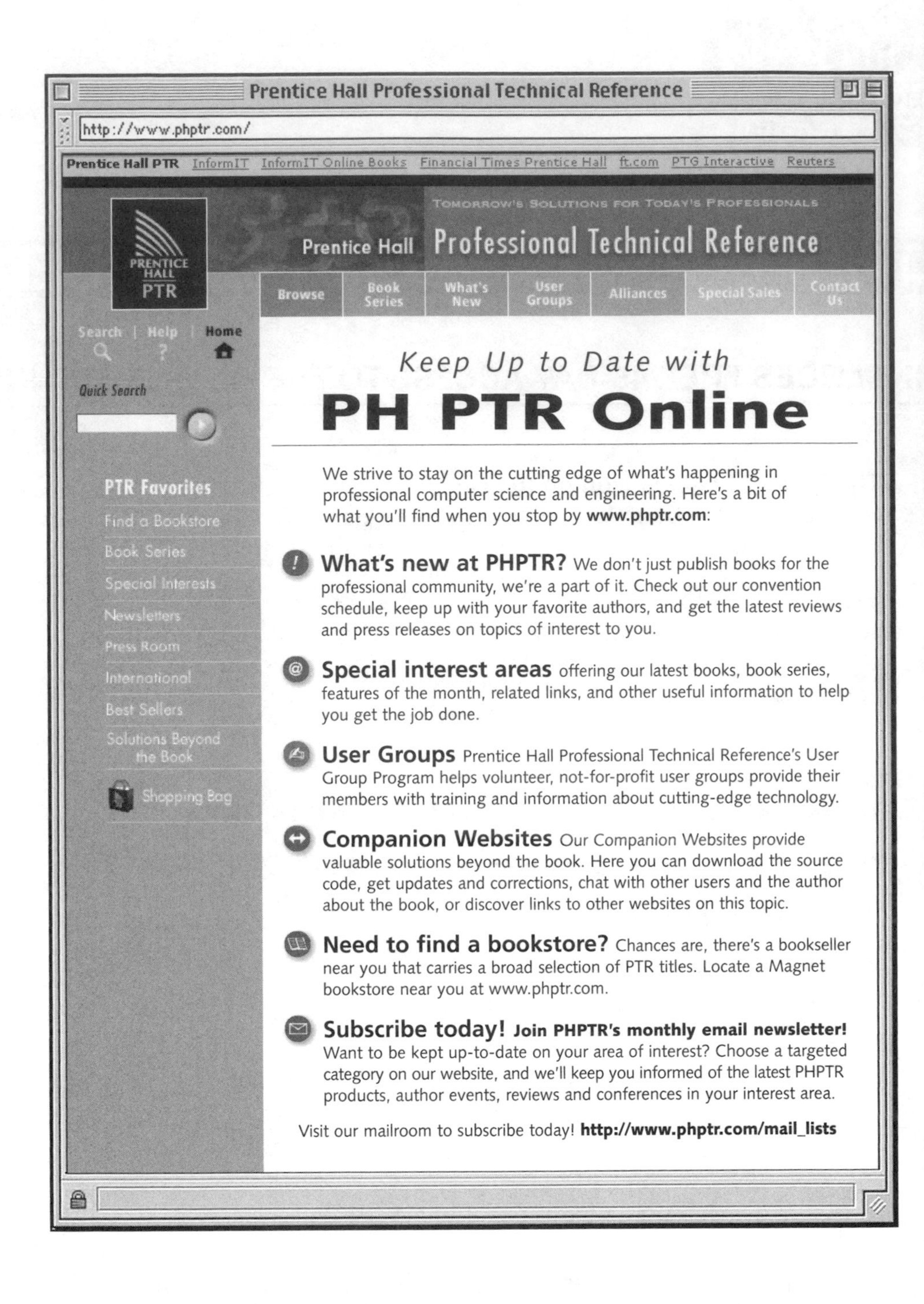

Prentice Hall Professional Technical Reference
http://www.phptr.com/
Prentice Hall PTR InformIT InformIT Online Books Financial Times Prentice Hall ft.com PTG Interactive Reuters
Tomorrow's Solutions for Today's Professionals
Prentice Hall Professional Technical Reference
PRENTICE HALL PTR
Browse
Book Series
What's New
User Groups
Alliances
Special Sales
Contact Us
Search | Help | Home
Quick Search
PTR Favorites
Find a Bookstore
Book Series
Special Interests
Newsletters
Press Room
International
Best Sellers
Solutions Beyond the Book
Shopping Bag
Keep Up to Date with
PH PTR Online
We strive to stay on the cutting edge of what's happening in professional computer science and engineering. Here's a bit of what you'll find when you stop by www.phptr.com:
What's new at PHPTR? We don't just publish books for the professional community, we're a part of it. Check out our convention schedule, keep up with your favorite authors, and get the latest reviews and press releases on topics of interest to you.
Special interest areas offering our latest books, book series, features of the month, related links, and other useful information to help you get the job done.
User Groups Prentice Hall Professional Technical Reference's User Group Program helps volunteer, not-for-profit user groups provide their members with training and information about cutting-edge technology.
Companion Websites Our Companion Websites provide valuable solutions beyond the book. Here you can download the source code, get updates and corrections, chat with other users and the author about the book, or discover links to other websites on this topic.
Need to find a bookstore? Chances are, there's a bookseller near you that carries a broad selection of PTR titles. Locate a Magnet bookstore near you at www.phptr.com.
Subscribe today! Join PHPTR's monthly email newsletter!
Want to be kept up-to-date on your area of interest? Choose a targeted category on our website, and we'll keep you informed of the latest PHPTR products, author events, reviews and conferences in your interest area.
Visit our mailroom to subscribe today! http://www.phptr.com/mail_lists